Quantum Chemistry

A Concise Introduction

Copyright © 2024 by Jeff Wang
All rights reserved. No part of this publication may be reproduced, distributed, or transmitted in any form or by any means, including photocopying, recording, or other electronic or mechanical methods, without the prior written permission of the publisher, except in the case of brief quotations embodied in critical reviews and certain other noncommercial uses permitted by copyright law.

Contents

1 Introduction to Quantum Mechanics and Chemistry   **11**
- 1.1 The Birth of Quantum Mechanics 11
- 1.2 Key Principles of Quantum Mechanics 13
- 1.3 Wave-Particle Duality 15
- 1.4 The Schrödinger Equation 16
- 1.5 Quantum Mechanics and Chemistry 19
- 1.6 Atomic Orbitals and Electron Behavior 21
- 1.7 Principles of Quantum Mechanics Applied to Chemistry   24
- 1.8 Interpreting Quantum Numbers in Chemistry 26
- 1.9 Quantum Tunneling and its Implications 28
- 1.10 Challenges and Limitations of Quantum Mechanics in Chemistry 29

2 Fundamentals of Quantum Chemistry   **33**
- 2.1 Introduction to the Quantum World 33
- 2.2 Postulates of Quantum Mechanics 35
- 2.3 Operators in Quantum Chemistry 37
- 2.4 Mathematical Tools for Quantum Chemistry 39
- 2.5 Particle in a Box 42
- 2.6 Harmonic Oscillator 44
- 2.7 Quantum States and Wavefunctions 46

2.8   Spin and Pauli Exclusion Principle   48

2.9   Hund's Rule and Atomic Term Symbols   50

2.10 Time-Dependent vs Time-Independent Schrödinger Equation   52

2.11 Normalization and Orthogonality in Wavefunctions .   54

3 Atomic Structure and Electronic Configuration   57

3.1   Overview of Atomic Structure   57

3.2   Nucleus: Protons and Neutrons   60

3.3   Electrons and Electron Clouds   62

3.4   Quantum Numbers and Electron Orbitals   64

3.5   Electron Configuration in Multi-electron Atoms . . .   66

3.6   The Aufbau Principle   67

3.7   Hund's Rule and Electron Filling   69

3.8   Periodicity in Electron Configuration   70

3.9   Ionization Energy and Electron Affinity   72

3.10 Shielding and Penetration Effect on Electron Configuration   75

3.11 The Role of Quantum Chemistry in Predicting Electron Configurations   77

4 Chemical Bonding and Molecular Structure   81

4.1   Introduction to Chemical Bonding   81

4.2   Ionic Bonding: Transfer of Electrons   83

4.3   Covalent Bonding: Sharing of Electrons   85

4.4   Metallic Bonding: Electron Sea Model   88

4.5   Molecular Orbital Theory   90

4.6   Hybridization in Covalent Bonds   92

4.7   Valence Shell Electron Pair Repulsion (VSEPR) Theory   94

4.8   Molecular Geometry and Polarity   96

4.9   Intermolecular Forces   99

- 4.10 Resonance Structures in Molecular Stability 101
- 4.11 Hydrogen Bonding and its Special Properties 103
- 4.12 Computational Predictions of Molecular Structures . 104

5 Quantum Mechanical Models and Approximations — 107
- 5.1 Overview of Quantum Mechanical Models 107
- 5.2 The Born-Oppenheimer Approximation 109
- 5.3 Perturbation Theory: Concept and Applications ... 111
- 5.4 Variational Principle and Method 113
- 5.5 Hartree-Fock Theory and Self-Consistent Field 115
- 5.6 Density Functional Theory (DFT) 117
- 5.7 Semi-empirical and Empirical Methods 119
- 5.8 Monte Carlo Simulations in Quantum Chemistry ... 121
- 5.9 Molecular Dynamics in Quantum Models 123
- 5.10 Advantages and Limitations of Computational Approximations 125

6 Computational Methods in Quantum Chemistry — 129
- 6.1 Introduction to Computational Quantum Chemistry . 129
- 6.2 Ab Initio Methods: Basis Sets and Electron Correlation 131
- 6.3 Semi-Empirical Methods: Parameters and Applications 133
- 6.4 Density Functional Theory (DFT): Foundations and Functional 136
- 6.5 Quantum Monte Carlo Methods 137
- 6.6 Molecular Mechanics Versus Quantum Mechanics .. 139
- 6.7 Software and Tools in Quantum Chemistry 142
- 6.8 Quantum Chemistry Simulations: Setup and Execution 144
- 6.9 Analyzing Quantum Chemistry Computational Results 146

7 Spectroscopy and Quantum Chemistry — 149
- 7.1 Introduction to Spectroscopy in Quantum Chemistry 149

	7.2	Principles of Spectroscopic Techniques	151
	7.3	Rotational Spectroscopy: Quantum Mechanical View	154
	7.4	Vibrational Spectroscopy: IR and Raman	156
	7.5	Electronic Spectroscopy: UV-visible	158
	7.6	Nuclear Magnetic Resonance (NMR) Spectroscopy . .	161
	7.7	Electron Spin Resonance (ESR) Spectroscopy	163
	7.8	X-ray Crystallography and Quantum Theory	164
	7.9	Spectroscopy Data Interpretation and Quantum Predictions	167
	7.10	Advanced Spectroscopic Techniques in Quantum Chemistry	170
	7.11	Applications of Spectroscopy in Quantum Chemistry	172
8	**Quantum Chemistry in Action: Case Studies and Applications**		**175**
	8.1	Introduction to Practical Applications of Quantum Chemistry	175
	8.2	Drug Design and Quantum Chemistry	177
	8.3	Material Science and Quantum Simulations	179
	8.4	Quantum Chemistry in Renewable Energy	181
	8.5	Catalysis and Reaction Mechanisms	184
	8.6	Environmental Chemistry and Pollution Control . . .	185
	8.7	Quantum Computing for Chemical Problems	187
	8.8	Quantum Dynamics in Biological Systems	189
	8.9	Forensic Science Applications	191
	8.10	Challenges in Applying Quantum Chemistry in Industry	193
9	**Recent Advances in Quantum Chemistry**		**197**
	9.1	Overview of Contemporary Quantum Chemistry . . .	197
	9.2	Progress in Computational Power and Algorithms . .	199
	9.3	Advancements in Density Functional Theory (DFT) .	202

9.4	Machine Learning and Artificial Intelligence in Quantum Chemistry	203
9.5	Quantum Computing for Chemical Simulations . . .	206
9.6	High-Performance Computing in Quantum Chemistry	208
9.7	Multi-scale Modeling Techniques	210
9.8	Time-Dependent Dynamics Simulations	212
9.9	New Functional and Basis Sets	214
9.10	Green Chemistry and Sustainable Approaches	215
9.11	Quantum Chemistry in Nanotechnology and Nanomedicine	217

CONTENTS

Preface

This book, *Quantum Chemistry: A Concise Introduction*, aims to provide a foundational understanding of the underlying principles and applications of quantum chemistry. As a discipline that marries quantum mechanics with chemical problems, quantum chemistry offers profound insights into molecular behavior, chemical reactions, and material properties at the atomic and subatomic levels.

Quantum chemistry is a challenging yet immensely rewarding field of study. It requires comprehension of both theoretical constructs and practical applications to fully grasp the chemical phenomena from a quantum perspective. Therefore, this book is meticulously structured to ease the learning process for beginners. It furnishes the reader with a comprehensive overview starting from the basics of quantum mechanics, building up through atomic structure, bonding theories, quantum approximations, and computational methods. This progression ensures that the foundational concepts are cemented before moving on to more complex topics.

Furthermore, the book delves into advanced subjects such as spectroscopy techniques, practical applications in current chemical and pharmaceutical industries, and innovations at the frontier of quantum chemistry research. Each chapter of this book is outlined to stand alone, offering detailed exposition of specific facets within quantum chemistry.

The intended audience for this book ranges from undergraduate students taking their first steps in quantum chemistry to graduate students looking for a text that succinctly explains complex concepts. It also serves as a reference for professionals who wish to revisit fundamental principles or explore new advances in the field.

By the end of this book, readers will have been equipped with both the theoretical frameworks and practical understanding necessary to navigate and contribute to the field of quantum chemistry. This book not only aims to educate but also to inspire the next generation of chemists, theorists, and innovators who will push the boundaries of what is possible in chemical research through the power of quantum mechanics.

Chapter 1

Introduction to Quantum Mechanics and Chemistry

This chapter sets the stage for understanding the integration of quantum mechanics into the realm of chemistry. It begins with a historical overview of quantum mechanics, illustrating its emergence and fundamental principles. Major concepts such as wave-particle duality, the Schrödinger equation, and quantum numbers are explained. Additionally, the chapter discusses how these quantum mechanical principles apply specifically to chemical contexts, providing the reader with a foundational comprehension of how atomic and molecular structures are interpreted through quantum mechanics.

1.1 The Birth of Quantum Mechanics

The conceptual underpinnings of quantum mechanics originated in the early 20th century, as a response to classical physics' inability to explain certain phenomena observed at microscopic scales. A pivotal moment in this development was Max Planck's resolution of the black-body radiation problem in 1900. Planck proposed that energy is not continuous, but rather quantized, being emitted or absorbed in discrete amounts called quanta. This was formulated mathematically as:

$$E = h\nu$$

where E is the energy of a quantum, ν is the frequency of the radiation, and h (Planck's constant) is a fundamental constant of nature, approximately 6.626×10^{-34} Js. Planck's quantization was revolutionary as it challenged the long-standing belief in the continuity of energy transitions and laid the groundwork for quantum theory.

Following Planck's discovery, Albert Einstein in 1905 introduced the concept of the photon, the quantum of light, to explain the photoelectric effect. Einstein proposed that light could be thought of as consisting of discrete packets of energy, which contradicted the classical wave theory of light. The energy of these photons was given by the same relation Planck derived, which explained why light of a frequency below a certain threshold was unable to eject electrons from a metal surface, regardless of its intensity. Einstein's contribution can be expressed as:

$$E_{\text{photon}} = h\nu$$

This assertion that light could exhibit particle-like properties in addition to wave-like characteristics was fundamental to the dawning understanding of wave-particle duality.

Further contributions came from Niels Bohr in 1913 through his model of the hydrogen atom, which integrated the quantized nature of electron orbits around the nucleus to explain atomic emission spectra. Bohr's model was crucial as it provided quantization conditions for the allowed orbits, described by:

$$mvr = n\hbar$$

where m is the electron mass, v its velocity, r the radius of the orbit, n is a positive integer (the principal quantum number), and $\hbar$ is the reduced Planck's constant ($h/2\pi$).

The refinement and expansion of quantum mechanics continued with the development of matrix mechanics by Werner Heisenberg in 1925 and wave mechanics by Erwin Schrödinger in 1926. Heisenberg's matrix formulation abstractly represented physical observables such as position and momentum as matrices operating in a discrete state space. Schrödinger's wave mechanics, contrarily, represented these observables using wave functions described by the Schrödinger equation:

$$i\hbar \frac{\partial}{\partial t} \Psi(x,t) = \hat{H} \Psi(x,t)$$

where $\Psi(x,t)$ is the wave function of the system, $\hat{H}$ is the Hamiltonian operator, and i represents the imaginary unit.

The formalism of quantum mechanics was further unified by Paul Dirac who, in 1928, formulated the Dirac equation, integrating quantum mechanics with special relativity and predicting the existence of antimatter. Dirac's notation and formalism also helped blend the matrix and wave approaches into a comprehensive theory, enhancing the mathematical rigor and scope of quantum mechanics.

Through these pivotal advancements, quantum mechanics emerged as a robust framework capable of explaining an array of physical phenomena at microscopic levels, challenging and extending beyond the limits of classical mechanics. This dramatic shift not only revolutionized our understanding of the physical world but also set the stage for technological advancements and opened new fields of research in chemistry, physics, and materials science.

1.2 Key Principles of Quantum Mechanics

Quantum mechanics, a fundamental theory in physics, describes the properties and behavior of matter and light at the atomic and subatomic levels. This section will elaborate on the key principles central to quantum mechanics, which are essential for the understanding of chemical phenomena at the molecular level.

Quantum States and Superposition

Quantum systems are described by their quantum states, which are represented mathematically by state vectors in a Hilbert space. The principle of superposition states that if a quantum system can be in multiple possible states, it can also be in a state that is a superposition of these states. This can be expressed mathematically as:

$$|\psi\rangle = c_1|\phi_1\rangle + c_2|\phi_2\rangle + \cdots + c_n|\phi_n\rangle$$

where $|\psi\rangle$ is the state vector of the system, $|\phi_i\rangle$ are the possible states, and c_i are the coefficients representing the probability amplitudes, which are complex numbers.

Quantization

Quantization is another fundamental principle where physical quantities, such as energy, angular momentum, and magnetic moment, take discrete values called quanta. This is in contrast to classical mechanics, where these quantities can vary continuously.

For example, the energy levels of an electron in a hydrogen atom are given by:

$$E_n = -\frac{13.6 \text{ eV}}{n^2}$$

where n is the principal quantum number, an integer. The quantized nature of these energy levels leads to the emission or absorption of photons with specific energies when electrons transition between levels.

Uncertainty Principle

The Heisenberg Uncertainty Principle is a fundamental limit to the precision with which certain pairs of physical properties, such as position x and momentum p, can be known simultaneously. It does not result from experimental imperfections, but from the very nature of quantum mechanics. The principle is quantitatively expressed as:

$$\Delta x \Delta p \geq \frac{\hbar}{2}$$

where Δx and Δp are the standard deviations of position and momentum, respectively, and $\hbar$ is the reduced Planck's constant.

Wave-Particle Duality

Wave-particle duality is a concept that proposes that every particle or quantum entity can exhibit both particle-like and wave-like properties. A clear manifestation of this principle is demonstrated through the double-slit experiment, where particles such as electrons show interference patterns, a property typical of waves.

Correspondence Principle

The correspondence principle, formulated by Niels Bohr, states that the behavior of systems described by quantum mechanics reproduces classical physics in the limit of large quantum numbers. This principle ensures that quantum mechanics is consistent with classical physics at macroscopic scales.

These key principles form the backbone of quantum mechanics and set a foundation for the complex phenomena observed in quantum chemistry. Understanding these principles is crucial for the study of molecular orbitals, reaction dynamics, and other chemical properties that are deeply rooted in quantum mechanical nature.

1.3 Wave-Particle Duality

The concept of wave-particle duality lies at the heart of quantum mechanics and predicates the behavior of subatomic particles like electrons and photons. Its inception pivots around the early 20th century experiments that challenged classical views of physics.

One of the seminal experiments that led to the acceptance of wave-particle duality was Thomas Young's double-slit experiment, initially demonstrating light's wave-like behavior. When monochromatic light passes through two closely spaced slits, it creates an interference pattern on a detection screen, indicative of wave behavior through constructive and destructive interference.

Let's consider the representation of wave interference in mathematical terms. Let $\psi_1(x)$ and $\psi_2(x)$ represent the wave functions associated with light passing through each slit. The total wave function $\Psi(x)$ at any point on the screen is the sum of these two wave functions:

$$\Psi(x) = \psi_1(x) + \psi_2(x).$$

The intensity pattern on the screen, which represents the probability density of photons striking at different locations, is then proportional to the square of the magnitude of the total wave function:

$$I(x) \propto |\Psi(x)|^2 = |\psi_1(x) + \psi_2(x)|^2.$$

Expanding this relation and using the properties of wave functions, we find

$$I(x) \propto |\psi_1(x)|^2 + |\psi_2(x)|^2 + 2\text{Re}[\psi_1(x)\psi_2^*(x)],$$

where $\text{Re}[\cdot]$ denotes the real part, and $\psi_2^*(x)$ is the complex conjugate of $\psi_2(x)$. The final term represents the interference term.

However, the perplexity of wave-particle duality emerged when similar interference patterns were observed even when particles like electrons, traditionally considered as particles, were used in such double-slit setups. If electrons are sent through the setup one at a time, each electron strikes a specific point on the detection screen, suggesting particle-like behavior. Yet, over time, these individual points build up to form an interference pattern, revealing the wave-like nature of electrons.

To resolve this, de Broglie proposed that all matter has wave-like properties with a wavelength λ given by

$$\lambda = \frac{h}{p},$$

where h is Planck's constant and p is the momentum of the particle. This de Broglie wavelength supports the observation that even particles with mass exhibit wave-like behaviors under certain conditions.

The mathematical implication of this wave-particle duality is foundational for the Schrödinger equation, which describes how the wave function of a quantum system evolves. It captures the essence of this duality through a wave function Ψ, a complex function whose modulus squared at a given point represents the probability density of finding a particle at that point.

For example, the time-independent Schrödinger equation for a particle in a potential $V(x)$ is

$$-\frac{\hbar^2}{2m}\frac{d^2\Psi}{dx^2} + V(x)\Psi = E\Psi,$$

where $\hbar$ is the reduced Planck's constant, m is the particle's mass, E is the energy, and Ψ on the left-hand side of the equation is equated with Ψ on the right side multiplied by E, symbolizing that this function encapsulates both the particle's wave and particle nature.

Thus, quantum mechanics introduces a paradigm where classical distinctions between waves and particles are no longer applicable. Subatomic entities exhibit behaviors characteristic of both waves and particles, depending on the experimental setup and the nature of observations made. This duality enriches our understanding of the physical and chemical properties of matter at the most fundamental level.

1.4 The Schrödinger Equation

The Schrödinger equation, named after physicist Erwin Schrödinger who formulated it in 1926, is a fundamental equation that describes how the quantum state of a physical system changes over time. It plays a pivotal role in quantum mechanics by providing a comprehensive description of the wave function of a system, which in turn encapsulates all the information about a particle's quantum state.

Formulation of the Equation

The time-dependent Schrödinger equation for a non-relativistic particle is expressed as:

$$i\hbar \frac{\partial \psi(\vec{r}, t)}{\partial t} = \hat{H} \psi(\vec{r}, t)$$

where i is the imaginary unit, $\hbar$ is the reduced Planck's constant ($\hbar = h/2\pi$ with h being the Planck constant), $\psi(\vec{r}, t)$ is the wave function of the system, $\vec{r}$ represents the spatial coordinates, t symbolizes time, and $\hat{H}$ is the Hamiltonian operator which represents the total energy of the system.

In situations involving only spatial variables, for systems not dependent on time, the time-independent Schrödinger equation is used:

$$\hat{H} \psi(\vec{r}) = E \psi(\vec{r})$$

Here, E stands for the energy eigenvalues associated with the stationary states of the system.

The Hamiltonian

The Hamiltonian in quantum mechanics is analogous to the total energy expression in classical mechanics, typically comprising kinetic energy (T) and potential energy (V) terms, formulated as:

$$\hat{H} = \hat{T} + \hat{V}$$

$$\hat{T} = -\frac{\hbar^2}{2m} \nabla^2$$

$$\hat{V} = V(\vec{r})$$

where ∇^2 is the Laplacian operator representing the sum of second spatial derivatives which account for the kinetic energy resulting from the particle's motion, and m is the mass of the particle. The potential energy $V(\vec{r})$ depends on the particle's position in a field.

Solution to the Equation

Solving the Schrödinger equation involves finding a function $\psi(\vec{r}, t)$ that satisfies the equation for given boundary conditions and potentials. These solutions are crucial as they provide the probability amplitudes for the presence of particles at different positions and times. The solutions, or wave functions, must be square-integrable over all space, ensuring their normalization:

$$\int |\psi(\vec{r})|^2 d^3r = 1$$

The square modulus $|\psi(\vec{r})|^2$ of the wave function corresponds to the probability density function for the position of a particle.

Importance in Chemistry

In chemistry, the Schrödinger equation is fundamental in describing the behavior of electrons in atoms and molecules. The solutions to the equation provide electron orbitals, which are the basis for understanding chemical bonds, reaction dynamics, and properties of materials. For instance, by analyzing electron densities, chemists can predict how atoms in molecules will interact, thus deducing molecular geometry, bond strengths, and chemical reactivity.

Example calculations often focus on the hydrogen atom, as it presents a relatively simple system that still demonstrates key quantum mechanical properties. By solving the Schrödinger equation for the hydrogen atom, we obtain the familiar s, p, d, f orbitals which determine most of the chemical properties of elements.

Numerical Methods

Due to the complexity of the Schrödinger equation, especially for many-body systems, numerical methods and approximations are often employed. Techniques such as the Hartree-Fock method and density functional theory (DFT) simplify the calculations by approximating the many-electron wave function, providing a pragmatic balance between accuracy and computational feasibility.

As we delve further into the implications and applications of the Schrödinger equation within chemistry, it becomes evident how quantum mechanics profoundly influences our understanding of molecular structure and reactivity. This equation not only highlights the wave-like behavior of matter at microscopic levels but also underscores the foundational framework of theoretical and computational chemistry.

1.5 Quantum Mechanics and Chemistry

Quantum mechanics, while originating in the realm of physical sciences, plays an instrumental role in explaining several crucial aspects of chemical systems. This integration is achieved through the analysis of electron behavior and bonding within and between atoms.

In this segment of the book, the transition between fundamental quantum mechanics and its chemical applications is detailed. Molecules and atoms are inherently quantum systems, with electrons, nuclei, and their interactions governed by quantum principles. For instance, the chemical bonds that hold atoms together in molecules result from electrons in shared orbitals, a description that emerges out of quantum mechanical calculations.

First, let's explore the atomic scale of matter. Atoms, the basic building blocks of molecules, possess a nucleus surrounded by electrons. The distribution and movement of these electrons around the nucleus are described not in terms of discrete orbits but rather through probability density functions, or orbitals calculated using the Schrödinger equation. For a hydrogen atom, which harbors only a single electron, the solution to the Schrödinger equation reveals different energy states the electron can occupy, each associated with a specific atomic orbital.

$$\hat{H}\psi = E\psi$$

Here, $\hat{H}$ represents the Hamiltonian operator signifying the total energy (kinetic + potential) of the system, ψ is the wave function of the state, and E is the associated energy eigenvalue.

At the molecular level, quantum mechanics aids in the analysis of molecular structures and interactions through molecular orbital theory. In essence, atomic orbitals of constituent atoms combine to form molecular orbitals. These molecular orbitals are acquired by solving the Schrödinger equation for the electrons in a molecule where the electron cloud no longer belongs to a single atom but extends over the molecule.

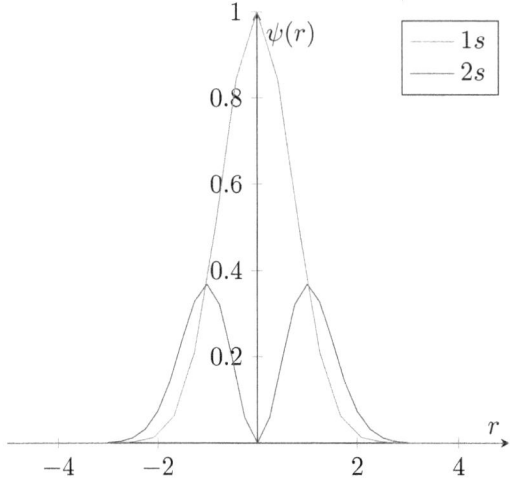

As illustrated above, the square of the wave function $\psi(r)$ at each point correlates to the electron density at that point, thus establishing a link between quantum states and observable chemical properties like bond order, bond length, and molecular reactivity. Electronegativity and bonding types are also derivable through these quantum considerations.

The Pauli exclusion principle, a quantum mechanical concept, fundamentally affects chemical behavior, ensuring that no two electrons in an atom can have identical quantum numbers. This principle is the origin of the electron configuration rules used in chemistry to explain the arrangement of electrons in an atom or molecule and consequently extends its implications to the periodic table structure and the nature of chemical bonding.

Photosynthesis and enzyme catalysis, quintessential chemical processes pivotal to life, involve electronic states and transitions, which can only be thoroughly elucidated using quantum mechanics. Photosynthesis, for instance, relies on the quantum tunneling of electrons - a purely quantum mechanical phenomenon - where electrons transfer through barriers that they classically shouldn't, as in the transition between excited states of chlorophyll during light absorption.

This interconnectedness of quantum mechanics and chemistry underscores the vital interplay between the abstract quantum world and tangible chemical phenomena, elucidating the complexity of molecular interactions, predicting chemical reactions, and discovering new materials.

Quantum mechanics provides a deep, fine-grained understanding of the basis of chemical properties and reactions. Each aspect, from atomic structure to molecular behavior under various conditions, relies heavily on quantum mechanics, illustrating that chemistry, at its core, is fundamentally a quantum science.

1.6 Atomic Orbitals and Electron Behavior

Atomic Orbitals

Atomic orbitals are fundamental to the understanding of chemical properties and reactions. An atomic orbital is a mathematical function describing the location and wave-like behavior of an electron within an atom. This function can be used to calculate the probability of finding an electron in any specific region around the atom's nucleus. The solutions to the Schrödinger equation for the hydrogen atom yield a set of quantum numbers and corresponding orbitals that exemplify the behavior of electrons in any atom.

Each orbital is characterized by three quantum numbers: the principal quantum number (n), the azimuthal quantum number (ℓ), and the magnetic quantum number (m_ℓ). The principal quantum number n corresponds to the electron's energy level and is a positive integer. The azimuthal quantum number ℓ describes the shape of the orbital and ranges from 0 to $n-1$. Each value of ℓ is associated with a specific type of orbital: $\ell = 0$ (s), $\ell = 1$ (p), $\ell = 2$ (d), $\ell = 3$ (f), etc. The magnetic quantum number m_ℓ, ranging from $-\ell$ to $+\ell$, denotes the orientation of the orbital in space.

Visualization of Orbitals

Using the hydrogen atom as an example, orbital visualizations are typically produced by plotting the probability density function $|\psi|^2$, where ψ is the wave function. The plot for $\psi_{1,0,0}$, representing the 1s orbital, shows a spherically symmetric "cloud" around the nucleus. Conversely, orbitals such as $\psi_{2,1,1}$ (2p orbital) indicate a more complex shape with nodal planes — regions where the probability of finding an electron is zero.

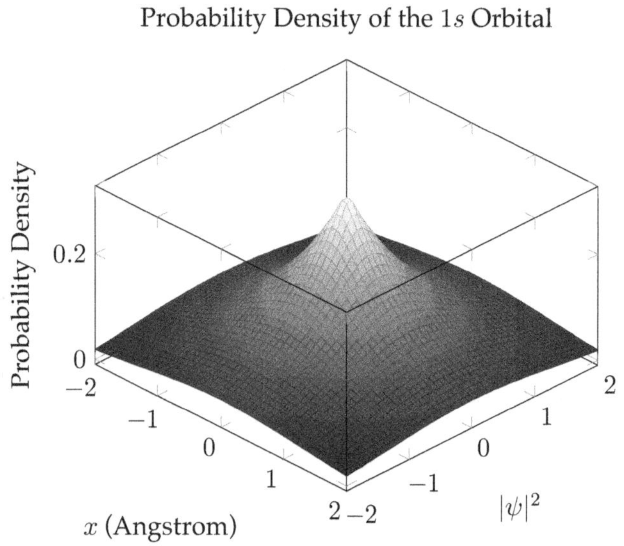

Figure 1.1: Visualization of the 1s orbital in a hydrogen atom.

Electron Behavior and Energy

Electrons in atoms exist in quantized energy states, determined by the values of n, ℓ, and m_ℓ. The lower n value, the lower the energy state and smaller the orbital, holding the electron closer to the nucleus. Transitions between these energy states occur via the absorption or emission of a photon, with energy corresponding to the difference between the initial and final states.

For multi-electron atoms, electron behavior becomes more complex due to electron-electron repulsions and the need for additional quantum numbers (such as spin quantum numbers). The Pauli exclusion principle dictates that no two electrons in the same atom can have identical sets of quantum numbers, which directly influences the electronic configuration and chemical characteristics of the atom.

Applicational Insight

In chemistry, understanding the interaction between atomic orbitals is crucial for predicting molecule behavior. Bond formation, for instance, involves the overlap of atomic orbitals. In the molecular orbital theory, atomic orbitals combine to form molecular orbitals, which are occupied by electrons shared between atoms. For example, the overlapping of two 1s orbitals from two hydrogen atoms can form bonding (lower energy) and antibonding (higher energy) molecular orbitals. This is represented by:

$$\text{HOMO: } \psi_{\text{bonding}} = \frac{1}{\sqrt{2}}(\psi_{1s, H1} + \psi_{1s, H2})$$

$$\text{LUMO: } \psi_{\text{antibonding}} = \frac{1}{\sqrt{2}}(\psi_{1s, H1} - \psi_{1s, H2})$$

Analyzing molecular orbitals gives a more comprehensive picture of the electron distribution within a molecule, enabling better understanding and prediction of chemical reactivity and properties.

Concluding Remarks

The study of atomic orbitals and electron behavior opens a window into the fundamental nature of chemical bonding and reactions. By applying quantum mechanics principles to electron dynamics, chemists can predict and explain a wide range of chemical phenomena, thereby extending the scope and accuracy of their experimental and theoretical frameworks.

1.7 Principles of Quantum Mechanics Applied to Chemistry

Quantum mechanics provides a fundamental framework for understanding the electronic structure of atoms and molecules, which is critical to the study of chemistry. The principles of quantum mechanics explain not only the behavior of atomic particles but also offer insights into chemical bonding, reaction dynamics, and material properties at the microscopic level.

One of the key aspects where quantum mechanics intersects significantly with chemistry is the concept of the atomic orbital. Atomic orbitals are solutions to the Schrödinger equation for electrons in atoms. These orbitals describe the probability distribution of electrons in an atom and are characterized by quantum numbers corresponding to the electron's energy, angular momentum, and magnetic orientation.

Atomic Orbitals and Electron Configuration

The Schrödinger equation, which is central to quantum mechanics, can be written as:

$$\hat{H}\psi = E\psi$$

where $\hat{H}$ is the Hamiltonian operator, ψ is the wave function of the system, and E is the energy eigenvalue associated with ψ. For atoms, solving the Schrödinger equation allows us to determine the energy levels and the shape of orbitals where electrons are likely to be found.

Each solution to the Schrödinger equation corresponds to a specific atomic orbital. Orbitals are often categorized by their quantum numbers: the principal quantum number (n), azimuthal quantum number (l), magnetic quantum number (m_l), and spin quantum number (m_s). These numbers not only describe the energy and shape of the orbital but also follow specific rules and limitations, crucial for the electron configuration in atoms.

For instance, let's consider the hydrogen atom, which is the simplest atom and a fundamental model in quantum chemistry. Its orbitals are determined by the principal quantum number n. The energy associated with each orbital is given by:

$$E_n = -\frac{13.6 \text{ eV}}{n^2}$$

1.7. PRINCIPLES OF QUANTUM MECHANICS APPLIED TO CHEMISTRY

where 13.6 eV is the Rydberg energy, the energy required to ionize the hydrogen atom from its ground state.

Chemical Bonding and Molecular Orbitals

Moving beyond single atoms, quantum mechanics is also essential in explaining the interactions that lead to chemical bonding. The formation of molecular orbitals can be described through linear combinations of atomic orbitals (LCAO). For example, in the hydrogen molecule (H_2), the molecular orbitals are formed by the constructive and destructive interference of the 1s atomic orbitals of individual hydrogen atoms.

The bonding molecular orbital (lower in energy) results in electron density between the two hydrogen nuclei, which can be represented as:

$$\psi_{bonding} = \frac{1}{\sqrt{2}}(\psi_{1sA} + \psi_{1sB})$$

where ψ_{1sA} and ψ_{1sB} are the wave functions of the 1s orbital of each hydrogen atom, respectively. This orbital leads to a net attraction between the two atoms, lowering the system's energy and resulting in a stable bond.

Conversely, the antibonding orbital (higher in energy) is formed by:

$$\psi_{antibonding} = \frac{1}{\sqrt{2}}(\psi_{1sA} - \psi_{1sB})$$

This results in a node, a region of zero electron density between the nuclei, leading to repulsion and an unstable arrangement if occupied.

Spectroscopy and Quantum Mechanics

Quantum mechanics also plays a pivotal role in spectroscopy, which is used to analyze the composition of substances. The interaction of electromagnetic radiation with matter can be understood via the quantization of energy levels in an atom or molecule. Absorption, emission, and scattering of photons involve transitions between these quantized energy states, each associated with specific wavelengths of light.

For example, in UV-Visible spectroscopy, the absorption of light corresponds to the excitation of electrons from the ground state to higher energy states. The wavelengths at which absorption occurs are directly related to the energy differences between these quantum states, providing a powerful method for chemical analysis and characterization.

Thus, understanding the principles of quantum mechanics is not only foundational for deciphering the atomic and molecular structure but is also imperative for advancing analytical techniques in chemistry, ultimately enhancing our ability to characterize and manipulate materials at the molecular level.

1.8 Interpreting Quantum Numbers in Chemistry

Quantum numbers are fundamental to the understanding of atomic structure in quantum chemistry. These numbers describe properties of the orbital where an electron resides, not the electron itself. The four quantum numbers — principal quantum number (n), azimuthal quantum number (l), magnetic quantum number (m_l), and spin quantum number (s) — define the size, shape, orientation, and spin of atomic orbitals, respectively. By deeply analyzing each of these quantum numbers, we can predict electron configuration, understand chemical bonding, and anticipate the magnetic and spectroscopic properties of atoms.

The principal quantum number, n, primarily determines the energy of an electron in a hydrogen-like atom and is a positive integer ($n = 1, 2, 3, \ldots$). For atoms with more than one electron, n also influences the orbital's size and the electron's distance from the nucleus. As n increases, the electron's average distance from the nucleus increases, and the electron spends more time further away from the nucleus, leading to higher energy levels and a greater potential for interaction with surrounding atoms or molecules.

Next is the azimuthal quantum number, l, which can take on any integer value from 0 to $n-1$ for each value of n. The number l determines the shape of the orbital, and it is associated with the angular momentum of the electron. Orbitals with different l values are known as s (sharp), p (principal), d (diffuse), and f (fundamental) orbitals corresponding to $l = 0, 1, 2, 3$, respectively. These shapes are critical for determining the type of chemical bonds an atom can form; for instance, the directional characteristics of p-orbitals are vital in the

1.8. INTERPRETING QUANTUM NUMBERS IN CHEMISTRY

formation of covalent bonds.

The magnetic quantum number, m_l, defines the orientation of the orbital in space relative to the other orbitals and can take integer values from $-l$ to $+l$, including zero. This quantum number becomes particularly important when considering the effects of external magnetic fields on the atom, as each orientation can respond differently, leading to phenomena such as Zeeman splitting.

Finally, the spin quantum number, s, has only two possible values, $+\frac{1}{2}$ and $-\frac{1}{2}$, corresponding to the two allowed directions of an electron's spin. The concept of spin is central in the explanation of the Pauli Exclusion Principle, which stipulates that no two electrons in an atom can have the same set of four quantum numbers, thus influencing the structure of the electron shells and chemical properties of the element.

Each set of quantum numbers for an electron in an atom is unique to that electron, thereby allowing chemists to predict electronic arrangements via the Aufbau principle, Hund's Rule, and the Pauli Exclusion Principle. These arrangements lead directly to predictions of chemical behavior, reactivity patterns, and even the physical properties of molecules such as their magnetic properties and optical activity.

To illustrate, consider the carbon atom, which has an electron configuration of $1s^2 2s^2 2p^2$. The electrons in the 2p orbitals have $n = 2$, $l = 1$, $m_l = -1, 0, 1$ (each electron in its own orbital as per Hund's Rule), and $s = +\frac{1}{2}$ or $-\frac{1}{2}$. The specific configuration and the unpaired electrons in the p orbitals allow carbon to form strong covalent bonds in organic chemistry, resulting in its ability to form a diverse array of complex molecules.

In summary, the interpretation of quantum numbers in chemistry provides not only a microscopic understanding of the electronic structure of atoms but also macroscopic insights into chemical reactivity and properties. This information is crucial for the synthesis of new materials, understanding of biological processes, and many other applications in science and engineering.

1.9 Quantum Tunneling and its Implications

Quantum tunneling is a quantum mechanical phenomenon where particles pass through a potential barrier that they classically should not be able to overcome due to insufficient kinetic energy. This behavior is one of the most striking consequences of the wave-like properties of particles as suggested by quantum mechanics and has profound implications in chemistry, particularly in the study of reaction mechanisms and enzyme activity.

To better understand quantum tunneling, consider a particle with energy E approaching a potential barrier of height V and width a. Classically, if $E < V$, the particle would be reflected by the barrier. However, in quantum mechanics, particles also exhibit wave-like properties, described by a wavefunction. The wavefunction can extend beyond the barrier, even if $E < V$, suggesting a probability of finding the particle on the other side of the barrier.

The phenomenon can be described mathematically by the Schrödinger equation:

$$-\frac{\hbar^2}{2m}\frac{d^2\psi}{dx^2} + V(x)\psi = E\psi$$

where ψ is the wavefunction of the particle, $\hbar$ is the reduced Planck's constant, m is the mass of the particle, $V(x)$ is the potential energy as a function of position x, and E is the total energy of the particle.

In regions where $E > V(x)$, the solution to the Schrödinger equation suggests that the particle behaves similarly to a free particle. However, in regions where $E < V(x)$, the wavefunction ψ decays exponentially, given by:

$$\psi(x) = \psi(0)e^{-\kappa x}$$

where $\kappa = \frac{\sqrt{2m(V-E)}}{\hbar}$.

The probability of the particle tunneling through the barrier, known as the transmission coefficient T, depends on the barrier's height and width, and can be estimated using the expression:

$$T \approx e^{-2\kappa a}$$

where a is the width of the barrier.

In chemical contexts, quantum tunneling is crucial in many low-temperature reactions where classical mechanics predicts that the reaction should not occur due to insufficient kinetic energy. This is particularly significant in enzymatic reactions and in the study of reaction rates. For instance, in hydrogen transfer reactions, which are common in various biochemical processes, tunneling allows hydrogen nuclei (protons) to 'tunnel' from one molecule to another, thus facilitating the reaction at a rate higher than would be possible classically.

Furthermore, tunneling has implications in modern technology. For example, it is a fundamental principle behind the scanning tunneling microscope (STM), which uses the tunneling of electrons from a sharp tip to a surface to produce images at the atomic level. STM has been instrumental in the study of surface chemistry and the development of materials sciences.

The incorporation of quantum tunneling in computational chemistry models also allows for better predictions of reaction mechanisms and pathways by providing insights that go beyond classical energy barriers. This includes the study of catalysis and the design of more efficient catalysts based on understanding how particles behave at the quantum level.

In summary, quantum tunneling extends beyond a mere quantum curiosity, influencing a broad range of chemical phenomena and applications. Its understanding not only enriches the knowledge of fundamental chemistry but also drives innovations in technology and industrial applications where chemical reactions play a crucial role.

1.10 Challenges and Limitations of Quantum Mechanics in Chemistry

Quantum mechanics, despite its substantial contributions to chemistry, presents several intrinsic challenges and limitations that curtail its application and interpretation in chemical contexts. We will delve into various facets of these limitations, focusing on computational constraints, the handling of large systems, non-covalent interactions, the quantum-classical boundary, and the treatment of electron correlation.

Computational Constraints: The solutions to the Schrödinger equation for many-electron atoms and molecules involve calculations that escalate in complexity and computational requirement with each additional electron and nucleus. This complexity stems from the need to account for every electron interacting with every other electron and nucleus, calculated through the wave function that is a function of the coordinates of all particles involved. As the size of the system increases, the computational cost grows exponentially.

For instance, solving the Schrödinger equation exactly for helium, the simplest many-body system after hydrogen, already requires considerable computational resources. Beyond systems with a few electrons, approximations are necessary. Techniques such as Hartree-Fock and Density Functional Theory (DFT) are utilized, which provide a balance between accuracy and computational feasibility. Even with these methods, limitations persist, particularly with very large molecules or densely packed materials.

$$\hat{H}\psi = E\psi$$

where $\hat{H}$ represents the Hamiltonian operator, ψ the wave function and E the energy eigenvalues. In larger systems, simplifying assumptions must be robust yet can introduce errors not present in smaller systems.

Handling of Large Systems: Quantum calculations on macroscopic systems are still out of reach with current technology and methodologies. Biological molecules like proteins, large polymers, or materials with extensive crystal lattices remain challenging to model accurately at the quantum level due to their size. Approximate methods like molecular mechanics or semi-empirical approaches such as PM3 often supplant pure quantum mechanical treatments in such systems. Quantum mechanics must often be integrated with classical physics methods to handle large scales effectively, leading to hybrid methods such as Quantum Mechanics/Molecular Mechanics (QM/MM) approaches.

Non-Covalent Interactions: The accurate representation of weak, non-covalent interactions (e.g., hydrogen bonding, van der Waals forces) is crucial for predicting structures, stabilities, and behaviors of molecular systems, particularly in biological contexts. Quantum mechanical methods straddle certain intrinsic discrepancies in describing these forces, often underestimating their contributions or requiring additional empirical corrections.

Quantum-Classical Boundary: The demarcation between quantum and classical descriptions (often referred to as the quantum-classical boundary) is not only a philosophical question but also a practical

1.10. CHALLENGES AND LIMITATIONS OF QUANTUM MECHANICS IN CHEMISTRY

Method	Strengths	Weaknesses
Hartree-Fock	Systematic	Fails to account for correlation
DFT	Efficiency	Struggles with dispersion forces
QM/MM	Scalable	Interface treatment complexity

Table 1.1: Comparison of various quantum computation methods

concern in simulations. Determining when a quantum description is necessary and when a classical approximation suffices is crucial for computational efficiency and accuracy. The transition from quantum behavior to classical behavior does not have a clear-cut boundary and remains a topic of ongoing investigation and debate.

Treatment of Electron Correlation: Electron correlation refers to the interaction between electrons in a quantum system, which is not adequately captured by the Hartree-Fock method where electrons are treated independently with a mean-field approach. Electron correlation is critical for accurately describing electronic structures, bond dissociation energies, reaction pathways, and spectroscopic properties. Advanced methods such as Configuration Interaction (CI) and Coupled-cluster (CC) are better at accounting for these effects but require significant computational effort.

While quantum mechanics offers a profound toolkit for understanding chemical phenomena, its full implementation in chemistry is hampered by computational limitations, the complexity of large molecular systems, inadequate treatment of non-covalent interactions, ambiguities at the quantum-classical interface, and challenges in accurately accounting for electron correlation. These limitations suggest areas for further research and development toward more efficient and comprehensive quantum mechanical methodologies.

Chapter 2

Fundamentals of Quantum Chemistry

This chapter delves into the core principles and theories that form the backbone of quantum chemistry. It covers the key postulates of quantum mechanics, introduces essential operators used in the field, and explores both the particle in a box model and the harmonic oscillator from a quantum perspective. The concept of quantum states and wavefunctions are explained, along with discussions on spin, Pauli's exclusion principle, and other foundational topics. These elements establish the essential framework needed to understand the subsequent computational and applied aspects of quantum chemistry.

2.1 Introduction to the Quantum World

Quantum chemistry is a branch of chemistry that investigates the behavior of atoms and molecules through the principles of quantum mechanics. The unique properties of atomic and molecular structures are assessed by theories that consider both particle-like and wave-like characteristics at subatomic scales.

Quantum mechanics emerged from the failure of classical physics to explain various atomic and molecular phenomena observed in the early 20th century. Among these, the discrete energy levels of electrons in an atom, as evident from atomic spectra, and the dual wave-particle nature of light and matter required a new theoretical framework.

One of the foundational experiments in quantum mechanics is the double-slit experiment. When particles such as electrons pass through two slits to a detection screen, they exhibit an interference pattern characteristic of waves. However, when observed closely, the particles hit the screen at discrete points, which is characteristic of particles. This duality underpins much of quantum chemistry.

Central to quantum mechanics are the concepts of quantization and the probabilistic nature of microscopic phenomena. Quantization refers to the existence of discrete energy levels rather than a continuum, as was traditionally assumed in classical mechanics. The probabilistic nature relates to the inability to simultaneously measure with perfect accuracy certain pairs of variables, like the position and momentum of a particle—known as the Heisenberg Uncertainty Principle.

The mathematical formulation of quantum mechanics is fundamentally based on the Schrödinger equation, an equation that describes how the quantum state of a physical system changes with time. It is expressed as:

$$i\hbar \frac{\partial}{\partial t} \Psi(\mathbf{r}, t) = \hat{H} \Psi(\mathbf{r}, t)$$

where i is the imaginary unit, $\hbar$ is the reduced Planck's constant, $\Psi(\mathbf{r}, t)$ is the wavefunction of the system, $\mathbf{r}$ denotes the position vector and t denotes the time, and $\hat{H}$ is the Hamiltonian operator representing the total energy of the system.

The wavefunction Ψ is a central concept in quantum mechanics, representing the state of a system at a given time. The square of the absolute value of the wavefunction, $|\Psi|^2$, is interpreted as a probability density function; it describes the probability distribution of a particle's position.

Quantum chemistry utilizes these principles to describe the electronic structure of atoms and molecules. By solving the Schrödinger equation for a molecular system, we can predict the distribution of electrons and thereby infer bond formation, molecular geometry, and properties like reactivity and spectroscopy. Quantum chemistry therefore plays a crucial role in fields such as material science, pharmacology, and nanotechnology, where understanding the electronic interactions at the atomic and molecular level is paramount.

In the subsequent sections, these foundational principles will be expanded upon to include a discussion of the specific operators used in quantum mechanics, the mathematical tools that facilitate the computation of these properties, and detailed examinations of simple quantum systems such as the particle in a box and the harmonic oscillator.

2.2 Postulates of Quantum Mechanics

Quantum mechanics is fundamentally built upon several key postulates which, though abstract, are essential for comprehending the behavior of systems at atomic and subatomic levels. These postulates form the structure upon which the mathematical formulation of quantum theory is hung. In the following paragraphs, we will discuss these postulates in a systematic approach, adhering strictly to standard formalisms used in quantum mechanics.

The State Postulate

The state of a quantum system is completely described by its wavefunction $\psi(\mathbf{r}, t)$, which contains all accessible information about the system. This wavefunction is a complex function, where $\mathbf{r}$ represents the spatial coordinates and t denotes time. The absolute square of the wavefunction, $|\psi(\mathbf{r}, t)|^2$, is proportional to the probability density of finding the particle at position $\mathbf{r}$ at time t. Mathematically, this is articulated as:

$$P(\mathbf{r}, t) = |\psi(\mathbf{r}, t)|^2,$$

where $P(\mathbf{r}, t)$ represents the probability density.

The Superposition Principle

The superposition principle states that if $\psi_1(\mathbf{r}, t)$ and $\psi_2(\mathbf{r}, t)$ are two wavefunctions that satisfy the Schrödinger equation, then any linear combination of these wavefunctions will also satisfy the Schrödinger equation. This can be expressed as:

$$\psi(\mathbf{r}, t) = c_1\psi_1(\mathbf{r}, t) + c_2\psi_2(\mathbf{r}, t),$$

where c_1 and c_2 are complex numbers. This principle enables the principle of interference and demonstrates the wave-like nature of quantum particles.

The Observable Postulate

In quantum mechanics, every observable quantity is associated with a Hermitian operator. The eigenvalues of this operator represent the possible outcomes of measurements of the observable. The operator corresponding to the measurement of a physical quantity A is denoted as $\hat{A}$, and its action on a wavefunction ψ that is an eigenvector can be described as:
$$\hat{A}\psi = a\psi,$$
where a represents a real-number eigenvalue. This formalism is critical for predicting measurement outcomes in quantum experiments.

The Measurement Postulate

The measurement postulate dictates that upon measuring a physical quantity, the system's wavefunction collapses to an eigenfunction of the corresponding operator. This implies that if a measurement of an observable $\hat{A}$ yields an eigenvalue a, the wavefunction of the system immediately after the measurement is ψ_a, the eigenvector of $\hat{A}$ related to the eigenvalue a. This process is inherently probabilistic, and the probability of collapsing to a particular eigenstate is given by the squared magnitude of the coefficient from the superposition of states.

The Time Evolution Postulate

The dynamic behavior of a quantum system is dictated by the Schrödinger equation, which is the central equation of motion in quantum mechanics. The time-dependent Schrödinger equation is expressed as:
$$i\hbar \frac{\partial \psi(\mathbf{r}, t)}{\partial t} = \hat{H}\psi(\mathbf{r}, t),$$
where $\hat{H}$ is the Hamiltonian operator of the system, representing the total energy (kinetic + potential), and i is the imaginary unit, and $\hbar$ is the reduced Planck's constant. This postulate is pivotal in predicting

how the quantum state evolves over time in response to interactions delineated by its Hamiltonian.

Despite their theoretical abstraction, these postulates provide a robust framework for analyzing a range of quantum phenomena and for developing advanced computational methods in quantum chemistry which underscore the subsequent sections of this text.

2.3 Operators in Quantum Chemistry

Operators are fundamental tools in quantum chemistry, playing a pivotal role in the manipulation and understanding of quantum states. Each operator corresponds to an observable physical property, such as momentum, position, or energy.

Definition and Overview of Operators

In quantum mechanics, an *operator* is defined as a mathematical entity that acts on the wavefunctions of the system, transforming them into other wavefunctions or into measurable values. One of the most central features of operators in quantum mechanics is that they are *linear*, meaning the operator acting on a linear combination of wavefunctions yields the same linear combination of the operator acting on the individual wavefunctions:

$$\hat{A}(c_1\psi_1 + c_2\psi_2) = c_1\hat{A}\psi_1 + c_2\hat{A}\psi_2$$

where $\hat{A}$ is an operator and c_1, c_2 are coefficients.

Hermitian Operators and Observables

In quantum mechanics, the measurable properties of a system are described by *Hermitian* operators (or self-adjoint operators). This class of operators is significant because their eigenvalues are real, which is a necessary feature for any quantity that is to be physically observable (e.g., position, momentum, energy).

A Hermitian operator $\hat{A}$ satisfies the relation:

$$\langle\psi|\hat{A}\phi\rangle = \langle\hat{A}\psi|\phi\rangle$$

for all wavefunctions ψ and ϕ. This relation ensures that the eigenvalues, which represent the measurable values of the observables, are real numbers.

Important Quantum Operators

Position and Momentum Operators: The position operator $\hat{x}$ in one dimension acts on a wavefunction $\psi(x)$ by multiplying it by x:

$$\hat{x}\psi(x) = x\psi(x)$$

The momentum operator in the position space, on the other hand, is given by:

$$\hat{p}_x = -i\hbar \frac{\partial}{\partial x}$$

where $\hbar$ is the reduced Planck's constant. When this operator acts on a wavefunction, it yields the momentum component along x.

Hamiltonian Operator: The most crucial operator in quantum mechanics is the Hamiltonian $\hat{H}$, representing the total energy of the system. For a single non-relativistic particle, it is given by the sum of kinetic and potential energy operators:

$$\hat{H} = \frac{\hat{p}^2}{2m} + V(\hat{x})$$

where $\hat{p}$ is the momentum operator, m is the particle's mass, and $V(\hat{x})$ is the potential energy as a function of position.

Commutation Relations

Commutation relations play a critical role in determining the properties of quantum systems. For two operators $\hat{A}$ and $\hat{B}$, the commutator is defined as:

$$[\hat{A}, \hat{B}] = \hat{A}\hat{B} - \hat{B}\hat{A}$$

A particularly important commutator is that between the position and momentum operators which encapsulates the fundamental nature of quantum mechanics:

$$[\hat{x}, \hat{p}_x] = i\hbar$$

This non-zero commutation signifies the inherent uncertainty in quantum mechanics, famously encapsulated in Heisenberg's uncertainty principle.

The operators in quantum chemistry are not just arbitrary mathematical tools; they are the very language through which the quantum mechanical description of the physical world is articulated. Each operator is meticulously linked to measurable properties and governs the transitions between different quantum states, underscoring their pivotal role in both the theory and application of quantum chemistry.

2.4 Mathematical Tools for Quantum Chemistry

The mathematical framework of quantum chemistry is predominantly built upon the principles of linear algebra, differential equations, and complex analysis. These mathematical tools are essential to understanding the behavior of systems at the quantum level, defining wavefunctions, operators, and other quantum observables.

Linear Algebra

At the heart of quantum mechanics are vector spaces, often complex Hilbert spaces, which support the principle of superposition of states. Key elements include:

- **Vectors** represent state functions of quantum systems. - **Operators** on these spaces represent observable properties such as momentum and energy. - **Eigenvalues and eigenvectors** of operators correspond to the measurable values of these properties and their associated states.

For a quantum state represented as a vector $|\psi\rangle$ in a Hilbert space, the application of an operator $\hat{A}$, which represents an observable, is given by:

$$\hat{A}|\psi\rangle = \lambda|\psi\rangle$$

where λ are the eigenvalues corresponding to the observable's measurable values.

Differential Equations

The Schrödinger equation, both time-independent and time-dependent forms, is a foundational differential equation in quantum mechanics. The time-independent Schrödinger equation is given by:

$$\hat{H}\psi(x) = E\psi(x)$$

where $\hat{H}$ is the Hamiltonian operator, $\psi(x)$ is the wavefunction of the state, and E is the energy eigenvalue associated with that state.

This equation is a type of eigenvalue problem in differential form and is typically solved to find the allowed energy levels and corresponding wavefunctions of quantum systems.

Complex Analysis

Since wavefunctions can have complex values, complex analysis provides the necessary tools to handle computations involving these functions. Complex conjugates and modulus are particularly important, as the probability density for finding a particle in a particular state is given by the modulus squared of the wavefunction:

$$\rho(x) = |\psi(x)|^2 = \psi(x)\psi^*(x)$$

where $\psi^*(x)$ is the complex conjugate of $\psi(x)$.

Matrices

In quantum chemistry, matrices are used extensively to represent operators, especially in systems with finite-dimensional state spaces. For example, the Hamiltonian, which encodes total energy information of the system, can be represented as a matrix in the basis of the system's eigenstates. The matrix representation of the Hamiltonian in a given basis $\{|i\rangle\}$ is:

$$\mathbf{H}_{ij} = \langle i|\hat{H}|j\rangle$$

Diagonalization of these matrices often yields the energy eigenvalues and the eigenvectors, corresponding to energy states and wavefunctions, respectively.

Fourier Transforms

The Fourier transform is a crucial tool in quantum mechanics for analyzing wavefunctions in both position and momentum spaces. For a wavefunction $\psi(x)$, its Fourier transform $\tilde{\psi}(p)$ in momentum space is given by:

$$\tilde{\psi}(p) = \frac{1}{\sqrt{2\pi\hbar}} \int_{-\infty}^{\infty} \psi(x) e^{-ipx/\hbar} dx$$

Dirac Notation

To handle abstract quantum states and operators formally and succinctly, Dirac notation is utilized. This bracket notation distinguishes between dual spaces (bra: $\langle\phi|$) and state spaces (ket: $|\psi\rangle$) effectively. Matrix elements, inner products, and outer products are compactly expressed as:

$$\langle\phi|\psi\rangle, \quad |\psi\rangle\langle\phi|$$

Tools for Computational Quantum Chemistry

Tools such as symbolic algebra systems and numerical solvers enable computation of complex quantum systems where analytical solutions are intractable. These computational tools handle large matrices, perform Fourier transforms, and solve differential equations numerically to simulate and predict quantum chemical behavior.

Mastery over these mathematical tools is indispensable in the pursuit of quantum chemistry, bridging the gap between abstract quantum theory and practical chemical applications.

2.5 Particle in a Box

The "particle in a box" model serves as an exemplary pedagogical tool to elucidate the fundamental principles of quantum mechanics. This model simplifies the complexities of a quantum system to a particle confined in a one-dimensional, infinitely high potential box. It is essential to understand this model as it introduces the concept of quantized energy levels, which are pivotal in quantum chemistry.

Model Definition

The particle in a box, also referred to as the infinite potential well, is characterized by a particle with mass m confined to a one-dimensional region from $x = 0$ to $x = L$, where L is the length of the box. The potential energy $V(x)$ within this model is defined by:

$$V(x) = \begin{cases} 0 & \text{if } 0 \leq x \leq L \\ \infty & \text{otherwise} \end{cases}$$

This condition of infinite potential means that the particle is reflected back into the box whenever it reaches the walls; therefore, it cannot exist outside the specified range.

Formulation Using the Schrödinger Equation

The time-independent Schrödinger equation for a one-dimensional system is given by:

$$-\frac{\hbar^2}{2m}\frac{d^2\psi}{dx^2} + V(x)\psi = E\psi$$

In regions where $V(x) = 0$, this equation reduces to:

$$-\frac{\hbar^2}{2m}\frac{d^2\psi}{dx^2} = E\psi$$

Given the boundary conditions $\psi(0) = \psi(L) = 0$ (as the wavefunction must drop to zero at the walls of the box where the potential is infinite), the general solution to this differential equation can be expressed in terms of sine and cosine functions. However, to satisfy

2.5. PARTICLE IN A BOX

the boundary conditions, the cosine terms are eliminated, and the solution simplifies to:

$$\psi_n(x) = A \sin\left(\frac{n\pi x}{L}\right)$$

where n is a positive integer, and A is the normalization constant.

Quantization of Energy

The quantization of energy emerges from substituting $\psi_n(x)$ back into the Schrödinger equation and solving for the energy eigenvalues:

$$-\frac{\hbar^2}{2m}\frac{d^2}{dx^2}\sin\left(\frac{n\pi x}{L}\right) = E_n \sin\left(\frac{n\pi x}{L}\right)$$

The second derivative of $\sin\left(\frac{n\pi x}{L}\right)$ is:

$$\frac{d^2}{dx^2}\sin\left(\frac{n\pi x}{L}\right) = -\frac{n^2\pi^2}{L^2}\sin\left(\frac{n\pi x}{L}\right)$$

This leads to the energy eigenvalues:

$$E_n = \frac{\hbar^2 n^2 \pi^2}{2mL^2}$$

where E_n indicates the energy associated with the n-th quantum state.

Normalization

The normalization constant A is determined by ensuring:

$$\int_0^L |\psi_n(x)|^2\, dx = 1$$

For $\psi_n(x) = A\sin\left(\frac{n\pi x}{L}\right)$, this condition transforms to:

$$A^2 \int_0^L \sin^2\left(\frac{n\pi x}{L}\right) dx = 1$$

The integral of $\sin^2$ over a single period is $L/2$, giving $A = \sqrt{\frac{2}{L}}$. Hence, the normalized wavefunctions:

$$\psi_n(x) = \sqrt{\frac{2}{L}}\sin\left(\frac{n\pi x}{L}\right)$$

Graphical Representation

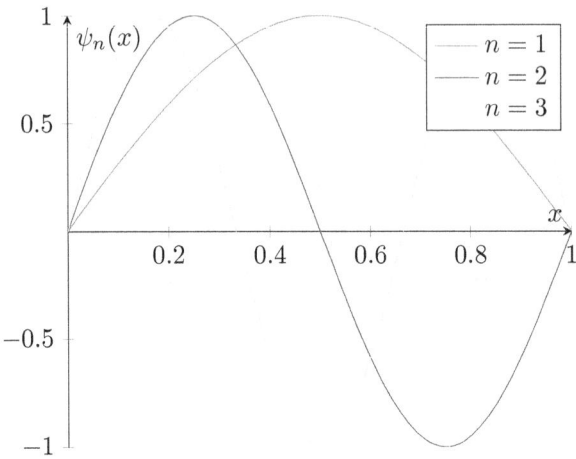

In summary, the particle in a box model is a cornerstone in quantum mechanics that introduces learners to the quantization of physical properties. It serves as a basis for understanding more complex quantum systems. The clear discrete energy levels and corresponding wavefunctions demonstrate the fundamental quantum mechanical principle that particles can only exist in certain allowed states, as indicated by their wavefunctions and energy levels.

2.6 Harmonic Oscillator

The quantum harmonic oscillator is a cornerstone model in quantum mechanics due to its extensive applicability across molecular vibrations, quantum field theory, and even in quantum optics. This model provides a pathway to understanding phenomena where particles exhibit motion around an equilibrium position.

The Model and Its Relevance

The harmonic oscillator model in quantum mechanics describes a particle of mass m attached to a spring with a force constant k, oscillating around an equilibrium position. Unlike the classical harmonic

2.6. HARMONIC OSCILLATOR

oscillator, where the particle can take any position with varying kinetic and potential energy, the quantum harmonic oscillator can only occupy specific quantized energy states.

The potential energy $V(x)$ of a harmonic oscillator is represented as:

$$V(x) = \frac{1}{2}kx^2$$

where x is the displacement from the equilibrium position.

The Schrödinger Equation for the Harmonic Oscillator

The time-independent Schrödinger equation for this system is given by:

$$-\frac{\hbar^2}{2m}\frac{d^2\psi(x)}{dx^2} + \frac{1}{2}kx^2\psi(x) = E\psi(x)$$

where $\hbar$ is the reduced Planck's constant, $\psi(x)$ is the wavefunction of the system, and E represents the energy levels of the system.

Solution to the Schrödinger Equation

The equation's solutions, $\psi_n(x)$, are derived using Hermite polynomials $H_n(\xi)$, where $\xi = \sqrt{\frac{m\omega}{\hbar}}x$ and $\omega = \sqrt{\frac{k}{m}}$ is the angular frequency of the oscillator. The eigenfunctions are given by:

$$\psi_n(x) = \left(\frac{m\omega}{\pi\hbar}\right)^{1/4} \frac{1}{\sqrt{2^n n!}} e^{-\xi^2/2} H_n(\xi)$$

These eigenfunctions are orthogonal, ensuring that each energy state is distinct and can be occupied by a particle independently of the others.

Energy Levels

The energy levels of the quantum harmonic oscillator are uniquely defined and equally spaced, a significant departure from potential wells where levels can cluster close to each other. Each level is given by:

$$E_n = \hbar\omega\left(n + \frac{1}{2}\right)$$

for $n = 0, 1, 2, \ldots$. The $\frac{1}{2}\hbar\omega$ term represents the zero-point energy, indicating that even in its lowest possible energy state (ground state), the oscillator exhibits non-zero energy due to the inherent uncertainty of quantum systems.

Ramifications and Physical Interpretation

The quantization of energy levels and the explicit forms of wave functions provide profound insights into the behavior of microscopic systems. One important application of the quantum harmonic oscillator is in describing the vibrational spectra of molecules. Since molecular vibrations often closely resemble the mechanics of a harmonic oscillator, this model allows for predictions regarding absorption spectra and hence is crucial for techniques like infrared spectroscopy.

Additionally, the harmonic oscillator model is instrumental in quantum field theory, where fields are treated analogously to oscillators at every point in space, leading to quantization of field states.

The quantum harmonic oscillator not only serves as a primary model for understanding various physical processes but also acts as a substantial foundational tool in advancing quantum mechanics towards complex and more applicable systems in chemistry and physics.

2.7 Quantum States and Wavefunctions

Quantum states represent the state of a system within the framework of quantum mechanics, encapsulating all information about the system. These states are typically described using wavefunctions, denoted by ψ, which are complex functions characterizing the probability amplitude for a particle or a system of particles in various positions and times. This section expounds on the principles underlying quantum states and wavefunctions, detailing their mathematical representations, physical implications, and constraints imposed by foundational principles of quantum mechanics.

The wavefunction $\psi(\mathbf{r}, t)$ is a function of spatial coordinates $\mathbf{r}$ and time t. The squared modulus of the wavefunction, $|\psi(\mathbf{r}, t)|^2$, is interpreted as the probability density that a particle is found at a certain position $\mathbf{r}$ at time t. Hence, for a wavefunction to serve as a valid physical representation, it must satisfy the normalization condition:

2.7. QUANTUM STATES AND WAVEFUNCTIONS

$$\int |\psi(\mathbf{r}, t)|^2 \, d\mathbf{r} = 1,$$

where the integral extends over all space, ensuring that the total probability of finding the particle somewhere in the space is one.

Furthermore, the wavefunction must be single-valued, continuous, and, in addition to its first derivative, must be continuous across all space. This requirement stems from the physical need for predictable and well-defined behavior across space and time.

In addressing the temporal aspect, quantum mechanics differentiates between time-dependent and time-independent scenarios. Time-independent wavefunctions, $\psi(\mathbf{r})$, are solutions to the Time-Independent Schrödinger Equation (TISE):

$$-\frac{\hbar^2}{2m}\nabla^2 \psi(\mathbf{r}) + V(\mathbf{r})\psi(\mathbf{r}) = E\psi(\mathbf{r}),$$

where $\hbar$ is the reduced Planck's constant, m is the particle's mass, ∇^2 is the Laplacian operator indicating the sum of second spatial derivatives, $V(\mathbf{r})$ is the potential energy function of the particle, and E is the energy eigenvalue associated with ψ. For stationary states, $\psi(\mathbf{r})$ provides the spatial component of the wavefunction.

As for time-dependent scenarios, introduced are solutions to the Time-Dependent Schrödinger Equation (TDSE):

$$i\hbar \frac{\partial}{\partial t}\psi(\mathbf{r}, t) = \hat{H}\psi(\mathbf{r}, t),$$

where $\hat{H}$ is the Hamiltonian operator, representing the total energy of the system. Solutions to the TDSE often involve separation of variables, yielding $\psi(\mathbf{r}, t) = \psi(\mathbf{r})\phi(t)$, where $\phi(t)$ encodes the temporal behavior, typically represented as $\phi(t) = e^{-iEt/\hbar}$, reflecting the wavelike nature of quantum mechanisms.

Given these foundational constructs, the notion of orthogonality and completeness in wavefunctions is significant in quantum mechanics. For a complete set of wavefunctions $\{\psi_n\}$ corresponding to a quantum system, any wavefunction ψ can be represented as a linear combination of these basis functions:

$$\psi = \sum_n c_n \psi_n,$$

where c_n are complex coefficients. The orthonormality condition for the wavefunctions, $\int \psi_m^*(\mathbf{r})\psi_n(\mathbf{r})\,d\mathbf{r} = \delta_{mn}$ (where δ_{mn} is the Kronecker delta and ψ_m^* is the complex conjugate of ψ_m), confirms that ψ_m and ψ_n are orthogonal for $m \neq n$.

In summary, the discourse on quantum states and wavefunctions provides a comprehensive understanding of how quantum systems are represented and studied in quantum mechanics. The mathematical rigor behind these representations underscores their pivotal role in predicting the behavior and interactions of particles at quantum scales, ultimately fostering deeper insights into the foundational underpinnings of quantum chemistry.

2.8 Spin and Pauli Exclusion Principle

Quantum chemistry delves deep into the intrinsic properties of particles at the atomic scale, properties that are not observable in classical mechanics. One such property is the quantum mechanical spin. Spin is an intrinsic form of angular momentum carried by elementary particles, composite particles (hadrons), and atomic nuclei.

Spin is quantized, meaning that it can only take certain discrete values. For electrons, which are fermions, the spin quantum number s is $\frac{1}{2}$. Consequently, the spin magnetic quantum number m_s can take values of $+\frac{1}{2}$ or $-\frac{1}{2}$, often referred to as "spin-up" and "spin-down", respectively.

In mathematical terms, spin is described by a set of spinors, which are solutions to the spin part of the quantum mechanical wave equation. The spin state of an electron is represented by a two-component spinor:

$$\chi_+ = \begin{pmatrix} 1 \\ 0 \end{pmatrix}, \quad \chi_- = \begin{pmatrix} 0 \\ 1 \end{pmatrix},$$

where χ_+ corresponds to spin-up and χ_- to spin-down.

Transitioning from the discussion of spin, we encounter the Pauli exclusion principle, a fundamental principle of quantum mechanics formulated by Wolfgang Pauli in 1925. According to this principle, no two fermions in a quantum system can have identical quantum

2.8. SPIN AND PAULI EXCLUSION PRINCIPLE

numbers. For electrons in an atom, this means that no two electrons can occupy the same quantum state simultaneously.

To illustrate this, consider the quantum numbers associated with electrons in an atom: the principal quantum number n, the azimuthal quantum number l, the magnetic quantum number m_l, and the spin quantum number m_s. The Pauli exclusion principle enforces that each electron in an atom must differ by at least one of these quantum numbers.

The exclusion principle is not just a theoretical concept; it has significant practical implications, influencing the structure of the periodic table and the behavior of atoms during chemical reactions. It explains the electron configurations and stabilities of different elements and is the reason for the formation of shells and subshells in atoms.

The principle can be visually represented in Hund's rules, which describe how electrons occupy subshells in atoms. Electrons will first fill unoccupied orbitals of the same energy (degenerate orbitals) before pairing up in any one orbital, and they do so by aligning their spins to maximize total spin, which results in the most stable configuration for an atom with incompletely filled shells.

The mathematical expression for the antisymmetry of the wavefunction of fermions due to the Pauli exclusion principle is represented as follows:

$$\Psi(x_1, x_2, \ldots, x_N) = \frac{1}{\sqrt{N!}} \sum_{\sigma \in S_N} (-1)^{\text{sgn}(\sigma)} \psi_{\sigma(1)}(x_1) \psi_{\sigma(2)}(x_2) \ldots \psi_{\sigma(N)}(x_N),$$

where N is the number of particles, S_N is the symmetric group on N elements, ψ_i are the wavefunctions, and $\text{sgn}(\sigma)$ denotes the sign of the permutation σ.

In summary, the introduction of spin and the Pauli exclusion principle into quantum chemistry provides a robust foundation for explaining many chemical phenomena. Their implications range from the arrangement of electrons in atoms to the structure and stability of molecules, highlighting the intricate link between quantum mechanics and chemical properties. These principles are fundamental in the study of atomic and molecular orbitals, influencing everything from simple diatomic molecules to complex biochemical structures.

2.9 Hund's Rule and Atomic Term Symbols

Hund's Rule

In quantum chemistry, Hund's rule is pivotal for predicting the electronic configuration of atoms, particularly in their ground states. It essentially states that electrons will fill degenerate orbitals in such a way as to maximize the number of electrons with the same spin. This rule arises from the Pauli exclusion principle and the Coulombic interactions between electrons.

When applying Hund's rule, the key points to consider are:

1. For a given electron shell, electrons occupy empty orbitals singly as far as possible.

2. Each electron in singly occupied orbitals has the same spin direction (either all ↑ or all ↓).

3. Pairing of electrons in orbitals only occurs when there are no vacant orbitals available in that subshell.

This arrangement reduces the repulsion between electrons, as maximally spaced electrons (occupying different orbitals) minimize the overlap of their probability density, thereby stabilizing the atom.

Atomic Term Symbols

Atomic term symbols provide a concise notation to describe the quantum states of electrons in atoms, particularly focusing on electron configurations contributing to the atom's energy state. Each term symbol is represented as $^{2S+1}L_J$, where:

- S represents the total spin quantum number, obtained by summing the spin quantum numbers of the individual electrons.

- L is the total orbital angular momentum quantum number. It is derived from summing the orbital angular momentum (ℓ) of individual electrons. The resultant L is denoted by a letter (e.g., S for $L = 0$, P for $L = 1$, D for $L = 2$, etc.).

- J is the total angular momentum quantum number that results from the vector coupling of L and S. It can range from $|L - S|$ to $L + S$ in integer steps.

2.9. HUND'S RULE AND ATOMIC TERM SYMBOLS

- The superscript $^{2S+1}$ indicates the multiplicity of the term, which relates to the possible orientation of the total spin.

An example that illustrates the use of atomic term symbols can be considered by looking at a carbon atom in its ground state. Its electronic configuration is $1s^2 2s^2 2p^2$. For the two electrons in the $2p$ orbitals:

- Each electron has a spin quantum number $s = \frac{1}{2}$.

- The $2p$ orbital implies $\ell = 1$ (each electron).

- They occupy different $2p$ orbitals following Hund's rule, maximizing the total spin $S = 1$.

- The possible values of L are obtained by considering the ways two p electrons can combine their angular momentum: resulting in $L = 0, 1, 2$ (denoted as S, P, D, respectively).

- Hence, the term symbols corresponding to these states are $^3S_1$, $^3P_2$, $^3P_1$, $^3P_0$, $^3D_3$, $^3D_2$, and $^3D_1$.

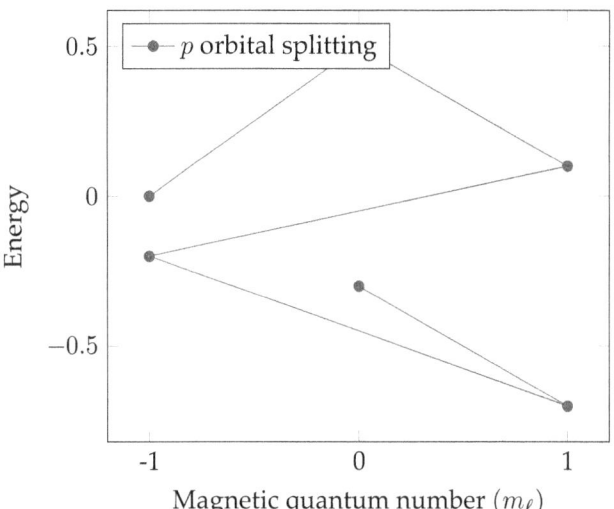

Term Splitting for $2p^2$ Configuration in Carbon

In summary, Hund's rule and atomic term symbols serve as essential tools for understanding and predicting the electronic structures and properties of atoms based on their electron configurations. These rules are crucial not just for academic study, but also for practical applications in spectroscopy, quantum computing, and other fields of chemistry and physics.

2.10 Time-Dependent vs Time-Independent Schrödinger Equation

The Schrödinger equation forms the cornerstone of quantum mechanics, providing the equation of motion for the quantum state of a physical system. It is generally formulated in two versions: the time-dependent Schrödinger equation (TDSE) and the time-independent Schrödinger equation (TISE). Understanding their differences and applications is crucial for grasping how quantum mechanics describes systems both dynamically and statically.

Time-Dependent Schrödinger Equation (TDSE)

The TDSE is used to determine the state of a quantum system that evolves over time. It is expressed as:

$$i\hbar \frac{\partial \Psi(\mathbf{r}, t)}{\partial t} = \hat{H} \Psi(\mathbf{r}, t),$$

where i is the imaginary unit, $\hbar$ is the reduced Planck's constant, $\Psi(\mathbf{r}, t)$ is the wavefunction of the system depending on position $\mathbf{r}$ and time t, and $\hat{H}$ is the Hamiltonian operator representing the total energy of the system.

This equation is pivotal when handling systems where the Hamiltonian itself is a function of time, or in scenarios like chemical reactions, where electronic and nuclear motions are interdependent and evolve. The TDSE underpins the dynamic simulation of molecules in fields such as femtochemistry.

Time-Independent Schrödinger Equation (TISE)

In contrast, the TISE is particularly effective in describing the stationary states of a system, where the Hamiltonian does not depend on time. The TISE can be derived from the TDSE by assuming a solution

2.10. TIME-DEPENDENT VS TIME-INDEPENDENT SCHRÖDINGER EQUATION

form:
$$\Psi(\mathbf{r}, t) = \psi(\mathbf{r})e^{-iEt/\hbar}.$$

Inserting this into the TDSE and cancelling common terms, we arrive at:
$$\hat{H}\psi(\mathbf{r}) = E\psi(\mathbf{r}),$$

where E is the energy eigenvalue associated with the eigenfunction $\psi(\mathbf{r})$. This form of the equation is used to determine the allowed energy levels of a quantum system and the corresponding wavefunctions.

Comparative Analysis

The choice between TDSE and TISE generally depends on the problem at hand. TDSE is indispensable in non-stationary contexts where the potential or external conditions change with time. Studies involving photon absorption, electron transfer, and time-resolved spectroscopy require the use of TDSE.

Meanwhile, TISE simplifies to solve for systems in stable states or those under static external fields, such as atoms in a molecule, crystalline structures, or particles in a potential well. Problems in solid-state physics, quantum chemistry, and materials science often involve TISE to find energy levels and state functions.

Illustrative Example

Consider a particle in a one-dimensional box with walls at $x = 0$ and $x = L$. Using TISE, we can determine its stationary states:

$$\psi_n(x) = \sqrt{\frac{2}{L}} \sin\left(\frac{n\pi x}{L}\right),$$

where n is a quantum number. For a TDSE perspective, if the width of the box L changes with time, the wavefunction needs to account for this time dependence, modifying the analysis.

Both TDSE and TISE are fundamental to describing quantum states, their use determined by whether the system's conditions are time-variant or time-invariant. Mastery of both forms allows for a comprehensive understanding of quantum mechanical phenomena across a wide range of applications in science and technology.

2.11 Normalization and Orthogonality in Wavefunctions

Wavefunctions form the cornerstone of our understanding of quantum systems, being central to the description of the state of particles in quantum mechanics. An essential aspect of dealing with wavefunctions is ensuring they are properly normalized and checking for their orthogonality when dealing with systems comprising multiple states or particles. This discourse will rigorously define normalization and orthogonality in the context of quantum wavefunctions, elucidate their significance, and then explore their applications in quantum mechanical problems.

Normalization is a process by which a wavefunction, $\psi(x)$, is scaled so that the total probability of finding the particle described by $\psi(x)$ in any region of its domain is 1. Mathematically, this is expressed via the integral over all space:

$$\int |\psi(x)|^2 \, dx = 1.$$

This condition ensures that the wavefunction is physically meaningful, as physical probabilities must total to one.

To fully understand why normalization is vital, consider a wavefunction that is not normalized. Its integral over all space might yield a value other than one, say C. If $C > 1$, this is non-physical as it implies a probability greater than 100% to find a particle somewhere in space. If $C < 1$, it suggests the particle is not fully accounted for in the described space. Thus, to render our mathematical model of the physical system meaningful, we adjust the wavefunction $\psi(x)$ by a factor so that the integral equals one:

$$\psi(x) \to \frac{1}{\sqrt{C}} \psi(x).$$

Orthogonality in wavefunctions pertains to the relationship between different wavefunctions describing independent quantum states of a system. The orthogonality condition for two different normalized wavefunctions, $\psi_i(x)$ and $\psi_j(x)$, is defined as:

$$\int \psi_i^*(x) \psi_j(x) \, dx = \delta_{ij},$$

2.11. NORMALIZATION AND ORTHOGONALITY IN WAVEFUNCTIONS

where $\psi_i^*(x)$ is the complex conjugate of $\psi_i(x)$, and δ_{ij} is the Kronecker delta, which is 1 if $i = j$ and 0 otherwise. This condition ensures that the quantum states are independent and physically distinct; that is, the measurement of one state gives no information about another.

Consider the instance of electrons in an atom, each described by a wavefunction. The orthogonality of these wavefunctions ensures that each electron occupies a distinct state, which aligns with the Pauli Exclusion Principle in multi-electron systems. Furthermore, in computational quantum chemistry, the orthogonality of basis functions (used to build molecular orbitals) simplifies many calculations, like those involving Hamiltonian matrix elements.

To provide a concrete example, take the simple case of the particle in a one-dimensional box with perfectly reflecting boundaries at $x = 0$ and $x = L$. The normalized wavefunctions for this system are:

$$\psi_n(x) = \sqrt{\frac{2}{L}} \sin\left(\frac{n\pi x}{L}\right), \quad n = 1, 2, 3, \ldots$$

It is straightforward to verify their normalization:

$$\int_0^L \left(\sqrt{\frac{2}{L}} \sin\left(\frac{n\pi x}{L}\right)\right)^2 dx = 1,$$

and their orthogonality:

$$\int_0^L \left(\sqrt{\frac{2}{L}} \sin\left(\frac{n\pi x}{L}\right)\right) \left(\sqrt{\frac{2}{L}} \sin\left(\frac{m\pi x}{L}\right)\right) dx = \delta_{nm},$$

where n and m are integers indicating different quantum states.

In summary, normalization ensures that wavefunctions represent proper probability densities, while orthogonality distinguishes independent quantum states within a quantum system. Together, these properties not only maintain the physical and mathematical robustness of quantum mechanical descriptions but also facilitate practical computations within quantum chemistry.

Chapter 3

Atomic Structure and Electronic Configuration

This chapter provides a comprehensive exploration of the structure of atoms and the arrangement of electrons within these fundamental units of matter. It starts with an overview of the nucleus, consisting of protons and neutrons, and expands to the electron cloud model, illustrating how electrons occupy specific energy levels and sublevels. Various quantum numbers are discussed to explain their roles in determining electron configurations and orbital shapes. Key principles such as the Aufbau principle and Hund's rule are detailed to demonstrate how they govern electron filling in atomic orbitals, ultimately influencing the chemical behavior of elements.

3.1 Overview of Atomic Structure

The fundamental understanding of the atomic structure is pivotal for the exploration of quantum chemistry. An atom consists of a central nucleus surrounded by a cloud of electrons. The nucleus, which contains the majority of an atom's mass, is composed of positively charged protons and neutral neutrons. Electrons, which are negatively charged particles, orbit the nucleus and are essential in defining the chemical properties of an element.

Composition of the Nucleus

The nucleus forms the core of an atom and is densely packed with protons and neutrons, collectively known as nucleons. The number of protons in the nucleus, referred to as the atomic number (Z), uniquely identifies an element. Neutrons, on the other hand, contribute predominantly to the mass of the atom but do not influence its chemical identity. The sum of the number of protons and neutrons gives the mass number (A) of the atom.

Mathematically, this can be expressed as:

$$A = Z + N$$

where N is the number of neutrons.

Electron Cloud Model

Electrons in an atom are not stationary; they move around the nucleus in regions known as atomic orbitals. Early models of the atom, such as the Rutherford model and the Bohr model, had offered initial insights but could not satisfactorily explain all atomic phenomenons, particularly the chemical behavior of atoms and their stability in multi-electron systems.

The electron cloud model, developed from the principles of quantum mechanics, provides a more accurate and detailed description of electron distributions around the nucleus. It describes the probability distribution of finding an electron in a certain region around the nucleus rather than a fixed path. This model introduces the concept of atomic orbitals as mathematical functions that describe these probability densities.

Energy Levels and Sublevels

Electrons in an atom occupy energy levels, which are broadly divided into principal energy levels, sublevels, and orbitals:

- **Principal energy levels** are often represented by the quantum number n and increase in energy as n increases. These levels are quantized, indicating that only certain values are allowed.

- **Sublevels** are contained within principal energy levels and are designated as s, p, d, or f. The energy and shape of the orbital

3.1. OVERVIEW OF ATOMIC STRUCTURE

associated with each sublevel vary, affecting how electrons are distributed within an atom.

- **Orbitals** are specific regions within sublevels where there is a high probability of finding an electron. Each orbital can hold a maximum of two electrons with opposite spins.

The distribution of electrons across these energy levels and sublevels is governed by fundamental principles such as the Pauli exclusion principle, the Aufbau principle, and Hund's rule, which will be elaborated upon in subsequent sections.

Quantification and Orbital Representations

To describe an electron in an atom, four quantum numbers are used:

1. The principal quantum number (n) indicates the energy level.

2. The azimuthal quantum number (l), related to the shape of the orbital.

3. The magnetic quantum number (m_l), which indicates the orientation of the orbital in space.

4. The spin quantum number (s), which describes the spin of the electron.

These quantum numbers not only help in describing the electronic structure of atoms but also facilitate the understanding of chemical properties and the periodicity observed in the elements of the periodic table.

Chemical Implications of Atomic Structure

The electronic structure of an atom directly influences its chemical reactivity and properties. Elements with similar electron configurations often show similar chemical behavior, a foundational concept in the organization of the periodic table. The detailed understanding of electron distributions enabled by quantum chemistry plays a crucial role in predicting the reactivity, bonding, and properties of elements.

The overview of atomic structure provides a foundation upon which deeper insights into computational models and predictive analyses in quantum chemistry are built. Subsequent sections will delve into the specifics of electron configurations and how quantum mechanics facilitates our understanding of complex atomic interactions and properties.

3.2 Nucleus: Protons and Neutrons

The nucleus, serving as the central core of an atom, is constituted predominantly of two types of subatomic particles: protons and neutrons, collectively known as nucleons. The nucleus plays a crucial role in defining the atomic and mass numbers of an atom, which in turn categorizes an element in the periodic table and influences its chemical properties.

Composition and Properties of Protons

Protons are positively charged particles with a charge of $+1e$ (where e is the elementary charge approximately equal to 1.602×10^{-19} coulombs). The mass of a proton is 1.672×10^{-27} kilograms, which is roughly equivalent to one atomic mass unit (amu). In the nucleus, the number of protons (denoted as Z) uniquely identifies the chemical element. For instance, carbon always has 6 protons.

Composition and Properties of Neutrons

Unlike protons, neutrons carry no net electric charge and are electrically neutral. However, they are slightly more massive than protons, with a mass of approximately 1.675×10^{-27} kilograms. The presence of neutrons contributes to the overall mass of the atom but does not affect its electrical charge. Neutrons play a crucial role in stabilizing the nucleus; without neutrons, the electrostatic repulsion between the positively charged protons would lead to nuclear instability.

3.2. NUCLEUS: PROTONS AND NEUTRONS

Binding Energy and Nuclear Forces

The particles in the nucleus are held together by a strong nuclear force, which acts over a very short range but is extremely powerful, overcoming the electrostatic repulsion between protons. The energy required to break a nucleus into its constituent protons and neutrons is known as the nuclear binding energy. It is indicative of the stability of a nucleus: the greater the binding energy, the more stable the nucleus.

Nuclear Models

To better understand the structure of the nucleus, physicists have developed models such as the liquid drop model and the shell model. The liquid drop model treats the nucleus like a drop of incompressible liquid, emphasizing the role of nuclear forces and surface tension in describing nuclear properties. The shell model, however, describes the nucleus in layers or shells of nucleons, similar to the arrangement of electrons in atoms. In this model, nucleons fill discrete nuclear shells, akin to electrons in atomic orbitals, which explains some aspects of nuclear stability observed in nature, particularly the existence of 'magic numbers' of protons and neutrons that correspond to exceptionally stable configurations.

Nucleus in Quantum Chemistry

In quantum chemistry, the behavior and interactions of nuclear particles are essential for understanding isotopes, nuclear reactions, and other phenomena that influence chemical behavior. Even though electrons mainly determine the chemical properties of elements, the nucleus significantly affects the isotopic identity of an element, thereby influencing its atomic mass and stability.

Isotopes and Isotonic Species

Isotopes are variants of a particular chemical element that have the same number of protons but different numbers of neutrons. For example, carbon-12 and carbon-14 are both isotopes of carbon, each with 6 protons, but carbon-12 has 6 neutrons whereas carbon-14 has

8 neutrons. This difference in neutron number can affect the stability, radioactive properties, and mass of isotopes, which in turn can influence chemical reactions and physical processes.

Similarly, isotonic species are nuclei that have the same number of neutrons but different numbers of protons. Isotones do not necessarily have similar chemical properties since they belong to different elements, but they exhibit similarities in nuclear properties.

Through this detailed exploration of the nucleus, we observe how intricately the properties of protons, neutrons, and nuclear forces interplay to dictate the behavior of atoms and elements in the broader context of quantum chemistry. This foundational knowledge is essential for further discussions on electron configuration and chemical properties as outlined in subsequent sections of this chapter.

3.3 Electrons and Electron Clouds

Electrons are subatomic particles with a fundamental negative charge and are considered one of the primary constituents of atoms. They play a crucial role in determining the chemical properties of elements and their interactions in various types of chemical bonds. Electrons have significantly lower mass compared to protons or neutrons, with a mass approximately 9.109×10^{-31} kilograms, or roughly 1/1836 the mass of a proton.

The behavior and arrangement of electrons in an atom are described by a model known as the electron cloud model. This model is rooted in quantum mechanics and replaces the earlier Bohr model. Unlike the Bohr model which depicted electrons traveling in fixed orbits around the nucleus, the electron cloud model describes electrons as being distributed within a region of space around the nucleus known as an "electron cloud". This cloud represents the probability density function of finding an electron at a particular location around the nucleus. Simply put, the density of the cloud at any given point correlates with the probability of an electron's presence at that point.

Wavefunction and Psi Squared: The position and state of an electron in an atom are described by a mathematical function known as the wavefunction (ψ). This function is complex-valued and its square modulus ($|\psi|^2$) represents the electron density at various points around the nucleus. The wavefunction solutions are derived from the Schrödinger equation, a fundamental equation in quantum

3.3. ELECTRONS AND ELECTRON CLOUDS

mechanics.

Orbitals and Shapes: Electron orbitals are specific regions of space where electrons are most likely to be found. These orbitals have various shapes (s, p, d, f) and sizes which are determined by quantum numbers associated with the electrons. The shape and size of an orbital affect the electron's energy level and its behavior in chemical bonding.

- The **s-orbitals** are spherical and centered around the nucleus.

- The **p-orbitals** are dumbbell-shaped and oriented along the x, y, and z axes.

- The **d-orbitals** and **f-orbitals** have more complex shapes, providing for a range of different spatial orientations and electron densities.

Quantum Mechanical Model and Energy States: In quantum mechanics, the energy states of electrons in an atom are quantized, meaning that they can only exist in certain discrete energy states. These states are defined by sets of quantum numbers describing unique conditions such as the principal quantum number (n), the azimuthal quantum number (l), the magnetic quantum number (m), and the spin quantum number (s). Each set of quantum numbers corresponds to a specific orbital and therefore to a specific distribution of electron density.

Electron Spin and Magnetic Properties: Electrons possess an intrinsic property known as spin, which contributes to their magnetic properties. Electron spin can be thought of as a form of angular momentum and can take one of two possible values, $\pm 1/2$. This property is critical in the formation of magnetic moments and the interpretation of spectroscopic data.

In summary, the electron cloud model provides a sophisticated description of electron positions surrounding the nucleus using quantum mechanics. Through illustrations of probability densities, shapes of orbitals, and quantum numbers, this model enhances our understanding of chemical bonding, atomic stability, and material properties.

3.4 Quantum Numbers and Electron Orbitals

Quantum numbers are fundamental to the understanding of atomic structure, specifically the arrangement and behavior of electrons in atoms. These numbers arise naturally from the solutions to Schrödinger's equation for the hydrogen atom, which demonstrates quantization within atomic systems. Each electron in an atom is described by a set of four quantum numbers: the principal quantum number (n), the azimuthal quantum number (l), the magnetic quantum number (m_l), and the spin quantum number (m_s).

Principal Quantum Number (n): The principal quantum number, n, determines the size and energy level of the orbital in which an electron resides. It can take any positive integer value ($n = 1, 2, 3, \ldots$). Higher values of n correspond to higher energy levels and larger orbitals. The energy of an electron in a hydrogen atom, neglecting electron-electron interactions in multi-electron atoms, is given by:

$$E_n = -\frac{Z^2 \cdot 13.6\,\text{eV}}{n^2}$$

where Z is the atomic number.

Azimuthal Quantum Number (l): The azimuthal quantum number, l, defines the shape of the orbital and subshell in which an electron can be found, and it is dependent on the value of n. The possible values of l are integers ranging from 0 to $n - 1$ for each value of n. Each value of l corresponds to a particular type of orbital: s (sharp; $l = 0$), p (principal; $l = 1$), d (diffuse; $l = 2$), and f (fundamental; $l = 3$).

l	Orbital Type	Shape
0	s	Spherical
1	p	Dumbbell-shaped
2	d	Cloverleaf-shaped
3	f	Complex shapes

Magnetic Quantum Number (m_l): The magnetic quantum number, m_l, specifies the orientation of the orbital in space relative to the other orbitals in the same subshell. m_l can have integer values between $-l$ and $+l$, inclusive. This quantum number is influenced by external magnetic fields and dictates the number of orbitals within a given subshell. For example, the p-orbitals, with $l = 1$, can have three possible values of m_l: -1, 0, and 1, resulting in three different p-orbitals (p_x, p_y, p_z).

Spin Quantum Number (m_s): The spin quantum number m_s describes the intrinsic spin of the electron, which is a fundamental property associated with its angular momentum. m_s can take one of two possible values: $+1/2$ or $-1/2$, indicating the two possible directions of electron spin (often conceptualized as "up" or "down").

When discussing electron orbitals, these quantum numbers are critical in determining the specific orbital where an electron can be found within an atom. For instance, the designation $1s$ indicates that the electron is in the first energy level ($n = 1$) and in an s orbital ($l = 0$). The complete description of an electron in this orbital would be $n = 1$, $l = 0$, $m_l = 0$, and $m_s = \pm 1/2$.

Visualizations often help in understanding complex concepts such as orbital shapes. The following representation uses TikZ to illustrate the shape of a p_z orbital:

Interactions among electrons and especially the exclusion principle necessitate a full understanding and application of these quantum numbers. This insight enables chemists and physicists to predict electron configuration, and subsequently, the chemical and physical properties of atoms.

3.5 Electron Configuration in Multi-electron Atoms

Multi-electron atoms are defined as atoms possessing more than one electron surrounding the nucleus. Configuring these electrons demands consideration of principles such as the Pauli exclusion principle, the Aufbau principle, and Hund's rule, often culminating in complex arrangements influenced by electron-electron interactions and orbital energies.

Pauli Exclusion Principle stipulates that no two electrons in an atom can have the exact same set of four quantum numbers. As such, an orbital can hold a maximum of two electrons, which must have opposite spins. The spin quantum number m_s is either $+1/2$ or $-1/2$.

Following this, the **Aufbau principle** guides the filling of orbitals with electrons. Orbitals are filled starting from the lowest energy level towards higher ones. This sequential filling is essential due to the stability contributed by lower energy configurations.

Hund's Rule states that electrons will fill an empty orbital in a given subshell before pairing up in one. This spread minimizes repulsion between electrons, which can significantly affect the atom's stability.

Electron Configurations Notation

The electron configuration of an atom is denoted by a distribution scheme:
$$(n)(\text{orbital type})^{(\text{number of electrons})}$$
where n represents the principal quantum number, the orbital type (s, p, d, f) speaks to the shape of the region in space occupied by the electron, and the superscript indicates the number of electrons in that orbital.

Sequence of Orbital Energies

The sequence in which orbitals are filled follows the order refined by observed electron fill patterns, which can sometimes deviate from the straightforward n+l rule due to inter-electron interactions and shielding. The following figure illustrates the generally accepted order.

3.6. THE AUFBAU PRINCIPLE

1s → 2s → 2p → 3s → 3p → 4s ↘ 3d → 4p → 5s ↘ 4d → 5p → 6s ↘ 4f ↘ 5d → 6p → 7s ↘ 5f ↘ 6d → 7p

Special Configurations and Anomalies

In some multi-electron atoms, such as those involving transition metals, the electron configuration may deviate from these predicted fill orders due to electron-electron repulsion and relative orbital energies. For instance, Chromium ($Z = 24$) is an example where instead of the expected configuration of $[Ar]3d^4 4s^2$, we observe $[Ar]3d^5 4s^1$. This anomaly arises from the extra stability gained by having a half-filled d-subshell.

Such exceptions emphasize the necessity of considering the actual energy of each specific configuration, which might not be accurately predictable by straightforward principles alone and can often require computational quantum chemistry methods for precise determination.

3.6 The Aufbau Principle

The Aufbau principle, derived from the German word "Aufbau" meaning "building up", forms the foundation for understanding the electron configurations of atoms in their ground states. It operates on the premise that electrons fill atomic orbitals in a sequence from the lowest energy level to higher energy levels, ensuring the most energetically favorable configuration.

Given the quantum mechanical nature of electrons, the Aufbau principle utilizes the energy levels defined by quantum numbers: principal quantum number (n), azimuthal quantum number (ℓ), magnetic quantum number (m_ℓ), and spin quantum number (m_s). Each set of these quantum numbers describes an orbital where electrons can reside, with energy primarily increasing with the principal quantum number (n). However, energy levels can also be influenced by the azimuthal quantum number which describes the shape of the orbital.

Typically, the order in which electrons fill these orbitals can be predicted using the ($n + \ell$) rule, where orbitals with a lower sum of n and ℓ are filled first. In cases where two orbitals have the same ($n+\ell$) value, the orbital with the lower n is filled first. This rule elucidates the electron configuration pattern seen across the periodic table and provides a primary framework before other rules, such as Hund's rule and Pauli's exclusion principle, are applied.

A simplified depiction of how electrons fill into orbitals can be mapped using electron configuration diagrams. A prime example is shown below through a TikZ diagram representing the filling order for hydrogen up to neon, illustrating the process of orbital filling.

```
1s ——+— 2 ——+——
      ↓       ↓
2s ——— 2 ——— 2p ——— 6 ———
                →
```

The arrows indicate the sequence of electron filling according to increasing energy levels asynchronously with the increase in atomic number.

Mathematically, the total energy of electrons in an atom can be approximated using the quantum numbers of the electrons. For hydrogen-like atoms, the energy of an electron in an orbital characterized by quantum numbers is given by the expression:

$$E_n = -\frac{13.6 \text{ eV}}{n^2}$$

where n is the principal quantum number. This equation indicates how energy becomes less negative, or higher, as n increases.

The practical application of the Aufbau principle in determining electron configurations inherently affects an element's chemical properties. For instance, knowing the electron configuration is vital for predicting chemical reactivity and bonding behavior, as seen in the variations across the periodic groups and periods.

Critically, while the Aufbau principle provides a starting point for establishing electronic configuration, exceptions to the filling order do occur, particularly due to electron-electron interactions and relativistic effects at higher atomic numbers. These anomalies need to be treated with other advanced quantum mechanical methods to accurately predict the electronic structure of more complex atoms.

The Aufbau principle is an essential tool in the arsenal of quantum chemistry for elucidating the underlying structure of atoms and their chemical characteristics. This fundamental rule aids both educators and students alike in navigating the complexities of electronic arrangements in differing elements.

3.7 Hund's Rule and Electron Filling

Hund's rule is a fundamental principle that guides the distribution of electrons among orbitals of the same energy level (degeneracy). It states that for a given electron shell, electrons fill unoccupied orbitals first before pairing with others in the same orbital. This behavior reduces electron-electron repulsion within the atom, allowing for a lower overall energy state. In order to visualize and comprehend this concept, a detailed analysis on the mechanism and implications of Hund's rule is imperative.

Electrons, as fermions, are subject to the Pauli exclusion principle, which requires that no two electrons in an atom can have the same set of four quantum numbers. The four quantum numbers — principal (n), angular momentum (l), magnetic (m_l), and spin (m_s) — define the energy, shape, orientation, and spin direction of an electron in an atom, respectively. According to Hund's rule, when multiple orbitals of the same energy are available (orbitals with the same n and l values but different m_l values), an electron occupies an empty orbital before it pairs up in the same orbital.

Consider filling electrons in the p orbitals of a carbon atom, for instance. Carbon has a total of six electrons. The electronic configuration notation for carbon can be examined up to the 2p level. The 2p subshell has three degenerate p orbitals (p_x, p_y, p_z, depicted with different m_l values). According to Hund's rule, the first three electrons occupy each of the three available p orbitals singly (one electron per orbital) and with parallel spins. This arrangement is illustrated using simple electron-in-box diagrams:

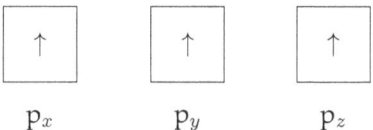

With three electrons and three p orbitals, each electron enters into an empty orbital according to Hund's rule, aligning their spins to minimize repulsion. Following this configuration, the next electron to be added to the carbon atom would pair up with one of the existing electrons in one of the p orbitals. The decision on which orbital the fourth electron pairs is arbitrary due to the equivalence in energy.

Hund's maximum multiplicity rule leads to the highest total spin state that an electron configuration can achieve, and thereby maximum stability, given the fact that electron repulsions in a singly occupied orbital are less than those in paired configurations. This higher total spin results in lower energy through exchange stabilization, a quantum mechanical effect that offers energy benefit for parallel spins in close proximity due to their symmetric wave function.

The implications of Hund's rule extend to a variety of phenomena in chemistry including the magnetic properties of atoms. Atoms that follow Hund's rule align their electron spins, potentially leading to paramagnetism when unpaired electrons are present. This principle also significantly affects the reactivity and bonding behavior of atoms.

Overall, Hund's rule, by dictating electron configurations in degenerate orbitals, plays a crucial role not only in determining the electronic structure of atoms but also in influencing their chemical characteristics. Understanding and applying this rule is essential in the prediction of electron configurations in multielectron atoms, which in turn determines the atom's chemophysical behavior.

3.8 Periodicity in Electron Configuration

In this section, we explore the periodicity in electron configuration, a fundamental principle that underpins the periodic table's structure and explains much about the chemical and physical properties of elements.

Understanding periodicity in electron configurations necessitates a detailed analysis of how electrons are distributed in atoms across different elements. This distribution is not arbitrary but follows a systematic pattern as outlined by the quantum mechanics principles covered in previous sections. The periodic trends in electron configuration arise due to the regular variations in atomic structure as one moves across or down the periodic table.

As elements in the periodic table are arranged in order of increasing atomic number, each successive element typically adds one proton to its nucleus and one electron to the electron cloud. The manner in which these additional electrons populate atomic orbitals is governed by previously discussed rules – the Aufbau principle, Pauli exclusion principle, and Hund's rule. Let's elucidate the effects of these principles on the periodic trends.

3.8. PERIODICITY IN ELECTRON CONFIGURATION

Role of Quantum Numbers: The quantum numbers of electrons—namely the principal quantum number (n), azimuthal quantum number (l), magnetic quantum number (m_l), and spin quantum number (m_s)—dictate the electron configuration of an atom. As n increases, so does the energy level of electrons and the size of their orbitals, leading to a higher energy and more spatially extended electron cloud.

Periodic Table Blocks and Electron Configuration: Electron configurations of elements can be systematically categorized according to the periodic table blocks (s, p, d, and f blocks) determined by the type of orbital (s, p, d, f) being filled with electrons:

- *s-block elements* (Groups 1 and 2 and Helium) have their outermost electrons in s orbitals.

- *p-block elements* (Groups 13 to 18) have their outermost electrons in p orbitals.

- *d-block elements* (Transition metals, Groups 3 to 12) have their outermost electrons in d orbitals, though the actual outermost electrons are in the s orbitals.

- *f-block elements* (Lanthanides and Actinides) feature electrons filling f orbitals.

Each of these blocks showcases a distinct trend in elemental properties, predominantly dictated by the electron configurations. For instance, within any single row (period), as we move from left to right, the electrons added to orbitals become more capable of shielding each other from the positive charge of the nucleus. This nuanced interplay of electron-electron repulsion, nuclear charge, and orbital shapes leads to periodicity in properties such as atomic radius, ionization energy, and electron affinity.

Ionization Energy and Electron Affinity: To delve deeper into periodic trends, let's consider ionization energy (the energy required to remove an electron) and electron affinity (the energy change when an electron is added). Both properties are deeply intertwined with electron configuration. Elements with a full or half-full valence shell tend to have higher ionization energies due to increased electron stability. Conversely, elements just one electron away from achieving a stable configuration tend to exhibit high electron affinities.

These principles help explain phenomena such as why noble gases possess remarkably high ionization energies—due to their complete valence shells—and why alkali metals have low ionization energies, facilitating the loss of their single valence electron to form cations.

Visualizing Periodicity with Electron Configurations: This visual representation charts atomic number against ionization energy, showcasing periodic trends.

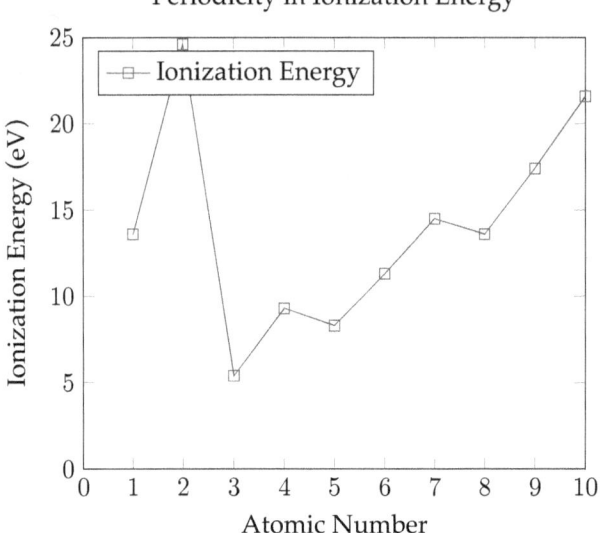

This graph not only highlights the stepwise increase in ionization energy across a period but also captures the dips at the start of a new period, reflecting a new electron shell being occupied.

Through detailed study of these periodic trends and underlying electronic structures, we gain indispensable insights into the behavior of elements, underscoring the predictive power of quantum chemistry.

3.9 Ionization Energy and Electron Affinity

Ionization energy (IE) and electron affinity (EA) are two fundamental concepts in quantum chemistry that are crucial for understanding the chemical reactivity and stability of atoms. These properties are indicative of the energy changes when electrons are either removed from or added to an atom or molecule.

Ionization Energy

Ionization energy is defined as the amount of energy required to remove an electron from an isolated gaseous atom or ion. The first ionization energy refers to the energy required to remove the first electron, while the second ionization energy is the energy required to remove the second electron, and so forth. Mathematically, the ionization energy can be represented for a generic atom (A) as follows:

$$A(g) \to A^+(g) + e^-; \quad \Delta H = IE_1$$

where IE_1 stands for the first ionization energy, ΔH denotes the change in enthalpy, and g signifies that the species are in the gaseous state.

Ionization energy is influenced by several factors including atomic size, nuclear charge, and electron shielding. Generally, ionization energy increases across a period from left to right in the periodic table due to the increase in nuclear charge. Conversely, it tends to decrease down a group as the outer electrons are further from the nucleus and more shielded by inner electrons.

Electron Affinity

Electron affinity is the amount of energy released when an electron is added to a neutral atom in the gas phase to form an anion. This process can be represented as:

$$A(g) + e^- \to A^-(g); \quad \Delta H = -EA$$

The EA value can be either positive or negative, depending on whether energy is released or absorbed when an electron is added. Atoms with a more negative EA are typically more eager to gain an electron.

Electron affinity also varies across the periodic table. It usually increases across a period due to the increasing nuclear charge which attracts the added electron more strongly. However, there are some exceptions due to electronic configuration.

Comparison and Trends

To represent these trends graphically, consider the following plots where the x-axis represents atomic number and the y-axis represents ionization energy and electron affinity respectively.

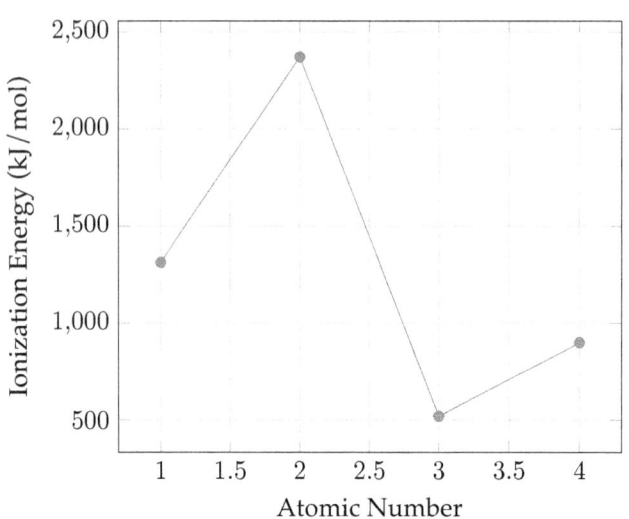

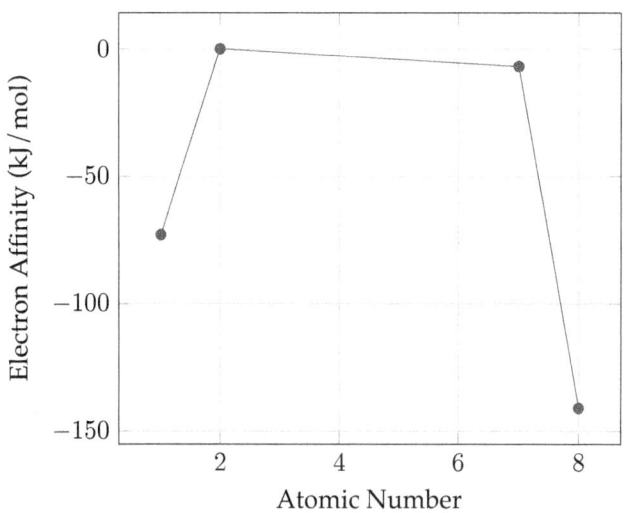

These graphs typify the general trends, though actual values might vary slightly due to experimental conditions and measurement accuracy. However, they clearly illustrate the periodic variations and anomalies in these key electron properties.

Chemical Implications

The understanding of IE and EA is crucial in predicting the chemical reactivity of elements. Elements with low ionization energies tend to be more metallic and are generally more reactive, readily losing electrons to form cations. Conversely, elements with high electron affinities are likely to gain electrons and form stable anions, showing nonmetallic characteristics.

In summary, ionization energy and electron affinity are pivotal in painting a comprehensive picture of electron behavior in atoms, influencing their chemical properties and interactions, thereby serving as a bedrock concept in the study of quantum chemistry.

3.10 Shielding and Penetration Effect on Electron Configuration

In understanding the electronic configuration of atoms, it is essential to consider not only the electron arrangement based on quantum numbers and principles stemming from the Pauli Exclusion Principle, Aufbau Principle, and Hund's Rule but also the effects of electron shielding and penetration. These concepts are crucial for explaining variations in atomic properties across the periodic table, such as ionization energy and electron affinity.

Shielding Effect

The shielding effect, also known as the screening effect, occurs when inner electrons act as a shield, reducing the effective nuclear charge (Z_{eff}) experienced by the outer electrons. The effective nuclear charge is the net positive charge exerted by the nucleus on an electron and is a critical factor in determining the electron's energy.

The effective nuclear charge can be quantitatively estimated using Slater's rules, which provide a mathematical approach to calculate the shielding constant σ:

$$Z_{\text{eff}} = Z - \sigma$$

where Z is the actual nuclear charge (i.e., the atomic number) and σ is the shielding constant determined by the distribution and number of intervening electrons.

Slater's rules involve assigning electrons to groups and then applying different shielding constants depending on the proximity of these groups to the electron in question. For example, electrons in the same quantum shell shield less effectively than electrons in closer inner shells. This difference in shielding leads to variations in Z_{eff} across different orbitals in the same shell or between different shells.

Penetration Effect

The penetration effect describes how some electron orbitals, such as s orbitals, have higher probabilities of being found close to the nucleus, even penetrating through the electron cloud of inner shells. This proximity to the nucleus means that electrons in s orbitals often experience a higher effective nuclear charge than electrons in other orbitals, such as p or d orbitals, which are generally more shielded by inner electrons.

To illustrate this, consider the electron configurations of sodium (Na) and magnesium (Mg). Both have neon (Ne) as their core configuration, but sodium has an additional $3s^1$ electron, whereas magnesium has a $3s^2$ configuration. Despite being in the same period (having the same principal quantum number), the s electrons in magnesium feel a significantly higher effective nuclear charge than sodium's s electron due to less effective shielding by the paired s electrons.

Implications on Electron Configuration

The cumulative effect of shielding and penetration leads to notable implications in the electron configuration and chemical properties of elements. One primary manifestation is the observation of irregularities in ionization energies and electron affinities. For instance, the unexpectedly low first ionization energy of aluminum compared to magnesium can be attributed to the additional p electron in aluminum, which is more shielded and less penetrating than the s electrons in magnesium.

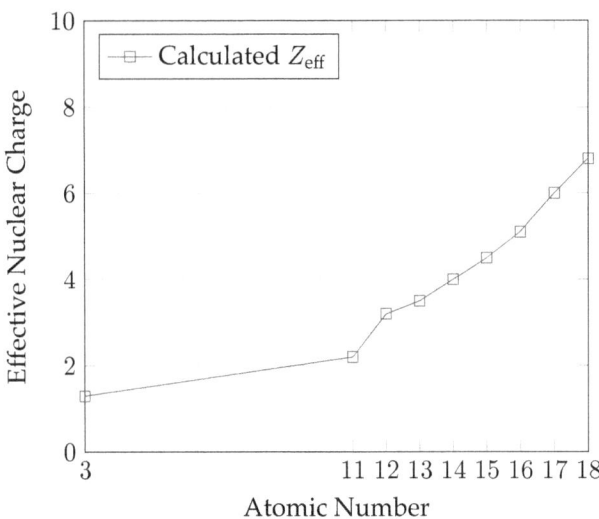

Effective Nuclear Charge Across a Period

The plot above models the increase in Z_{eff} as evidenced by more penetrating s electrons compared to p electrons which have more diffuse probability distributions away from the nucleus, reflecting reduced Z_{eff}. As a result, the observed chemical reactivity and trends in the periodic table can be closely aligned with these underlying electronic effects, adhering to quantum mechanical principles.

This detailed explanation elucidates how shielding and penetration effects modify the effective nuclear charge, impacting electron configurations and, consequently, the elemental behavior of atoms.

3.11 The Role of Quantum Chemistry in Predicting Electron Configurations

Quantum chemistry stands as a pivotal sub-discipline of chemistry, predominantly concerned with the application of quantum mechanics to chemical problems. It provides the theoretical framework to comprehend and predict the arrangement of electrons in atoms and molecules, a cornerstone concept known as electron configuration. This insight is crucial for elucidating the physical and chemical properties of substances, enabling chemists to rationalize reactivities, bonding patterns, and even material properties.

To predict electron configurations using quantum chemistry, one must begin with the Schrödinger equation, the fundamental equation of quantum mechanics for non-relativistic particles. For an atom, the Schrödinger equation is expressed as:

$$\hat{H}\Psi = E\Psi$$

where $\hat{H}$ is the Hamiltonian operator representing the total energy of the system, Ψ is the wave function of the electron, and E is the corresponding energy eigenvalue. Solving this equation facilitates the determination of possible energy levels (eigenvalues) and their associated wavefunctions (eigenstates) for electrons in atoms.

Firstly, quantum numbers arise naturally from the solutions to the Schrödinger equation in three-dimensional space, which dictate the distribution and energies of electrons. These are:

- Principal quantum number, n, which indicates the energy level and approximate radial distance of the electron from the nucleus.

- Angular momentum quantum number, l, symbolizing the shape of the orbital and having a maximum value of $n - 1$.

- Magnetic quantum number, m_l, defining the orientation of the orbital in space, which ranges from $-l$ to l.

- Spin quantum number, s, representing the two possible orientations of the electron's spin, either $+\frac{1}{2}$ or $-\frac{1}{2}$.

The solutions to the Schrödinger equation also support the visualization of atomic orbitals—the probable regions in space where electrons can be located. These orbitals form the basis for understanding electron configurations in multi-electron atoms, where the Pauli exclusion principle and electron-electron interactions come into play. The exclusion principle asserts that no two electrons in an atom can have the same set of four quantum numbers, thereby influencing the arrangement of electrons in orbitals.

With a foundational understanding of single atoms, quantum chemistry extends these principles to more complex systems, including molecules. Here, methods such as Hartree-Fock, Density Functional Theory (DFT), and post-Hartree-Fock are employed to solve the

3.11. THE ROLE OF QUANTUM CHEMISTRY IN PREDICTING ELECTRON CONFIGURATIONS

Schrödinger equation for systems with multiple electrons. These approaches approximate the many-body problem by considering electron densities or using trial wave functions, which incorporate the correlation and exchange of electrons.

For example, the usage of Slater determinants in the Hartree-Fock method constructs a wave function that satisfies the antisymmetry requirement (arising from the Pauli exclusion principle), predicting energy states and electron configurations more accurately for multi-electron systems. In more complex scenarios, DFT potentially provides a better approximation by using functionals of the electron density rather than wavefunctions, which simplifies calculations and is extensively used for larger systems.

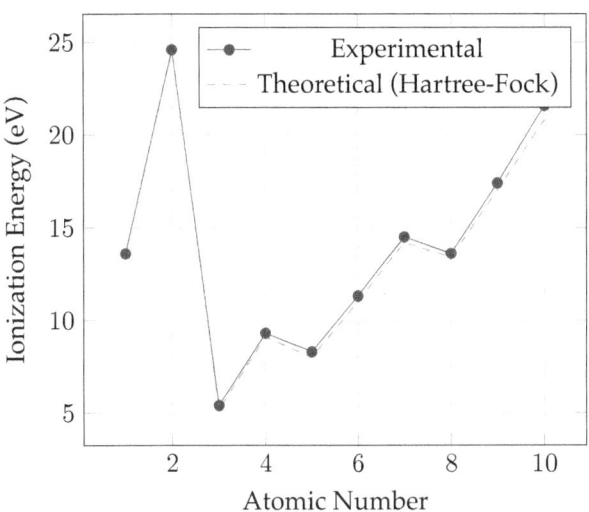

Figure 3.1: This plot demonstrates the effectiveness of Hartree-Fock calculations in predicting the ionization energies comparing to experimental values.

To evaluate the performance and accuracy of these quantum mechanical models, one can compare calculated properties (such as ionization energies, electron affinities) with experimental findings. Figure 3.1 illustrates a comparison of theoretical predictions by the Hartree-Fock method against experimentally measured ionization energies across several elements.

Quantum chemistry not only facilitates understanding and predicting electron configurations but is instrumental in bridging theoretical predictions with practical chemical phenomena, underscoring the profound implications of quantum mechanics in explaining and predicting the behavior of atomic and molecular systems.

Chapter 4

Chemical Bonding and Molecular Structure

This chapter examines the types of chemical bonds and their influence on molecular structure, crucial for understanding how compounds are formed and how they function. It presents a detailed discussion on ionic, covalent, and metallic bonding, alongside the theory of molecular orbitals and hybridization. The chapter further explores the Valence Shell Electron Pair Repulsion (VSEPR) theory to predict molecular geometry and delves into the significance of intermolecular forces in governing physical properties of substances. This foundational knowledge is essential for comprehending chemical reactivity and properties at molecular levels.

4.1 Introduction to Chemical Bonding

Chemical bonding is the process by which atoms combine to form new substances. The concept rests on the principle of electron interactions, which are the primary drivers in the formation of chemical bonds. Atoms bond to achieve more stable electronic configurations, typically aiming to fill their valence shells, which are the outermost electron shells in an atom. To grasp the nuances of chemical bonding, one must delve into the nature of these electron interactions and the mechanisms through which atoms can share, donate, or pool their electrons.

CHAPTER 4. CHEMICAL BONDING AND MOLECULAR STRUCTURE

The types of chemical bonds primarily encompass ionic, covalent, and metallic bonds, each differentiated by their method of electron management and interaction between participating atoms.

1. Ionic Bonding: Occurs when electrons are transferred completely from one atom to another, leading to the formation of positively charged ions (cations) and negatively charged ions (anions). This transfer typically happens between a metal and a non-metal. The resulting electrostatic attraction between oppositely charged ions forms the ionic bond. For example, in sodium chloride (NaCl), sodium donates an electron to chlorine, resulting in Na^+ and Cl^- ions.

2. Covalent Bonding: Involves the sharing of electrons between two atoms, typically non-metals, allowing each atom to gain electrons to complete its valence shell. This type of bond can be single, double, or triple, depending on the number of shared electron pairs. Covalent bonds are depicted using Lewis structures, which illustrate the arrangement of electrons in a molecule. For instance, in a water (H_2O) molecule, the oxygen atom shares electrons with two hydrogen atoms, forming two single covalent bonds.

3. Metallic Bonding: Described through the electron sea model, involves the pooling of valence electrons by metal atoms. These electrons are not associated with any specific atom and can move relatively freely through the metal lattice. This delocalization of electrons provides metals with unique properties such as conductivity and malleability.

In understanding these bonding types, it becomes evident why atoms select particular bonding strategies, governed by factors such as electronegativity (the ability of an atom to attract electrons), atomic size, and ionization energy. The electron configuration of an atom fundamentally influences the type and strength of the bonds it can form.

For quantitative analysis, the energy changes accompanying bond formation are crucial:

$$\Delta H = H_{products} - H_{reactants}$$

where ΔH is the change in enthalpy (heat energy associated with the bond formation). A negative ΔH indicates an exothermic reaction, where bond formation releases energy, suggesting a stable chemical compound.

Further elaboration on each type of bond will follow this introductory section, along with graphical renditions in subsequent chapters using tools such as molecular orbital theory and hybridization models, providing a deeper understanding of how atoms share or transfer electrons to form stable chemical structures.

Examples and illustrations in the following sections will use a variety of atom models and molecular representations, including ball-and-stick models, space-filling models, and energy-level diagrams, to aid in the visualization and understanding of complex molecular structures and bonding interactions. These models help convey the spatial arrangements of atoms within a molecule and the corresponding distribution of electrons, as fundamental to the prediction of molecular behavior in chemical reactions.

The foundational theories, such as the Octet Rule, which postulates that atoms are generally most stable when they possess eight electrons in their valence shell, will be integral throughout our exploration of chemical bonding.

With this background established, the next sections will dissect specific bonding types in greater detail, providing the chemical foundation upon which the structure and reactivity of molecules are understood and predicted.

4.2 Ionic Bonding: Transfer of Electrons

Ionic bonding represents a fundamental type of chemical bond that involves the electrostatic attraction between oppositely charged ions. This bond forms when one atom, typically a metal, loses one or more electrons, becoming a positively charged ion or cation, and another atom, typically a non-metal, gains those electrons, becoming a negatively charged ion or anion. The resultant electrostatic attraction between the oppositely charged ions leads to the formation of an ionic compound.

Electron Transfer and Formation of Ions

Atoms strive to achieve a stable electron configuration, often similar to that of the nearest noble gas. This tendency drives the transfer of electrons from metals, which have relatively low ionization energies, to non-metals, which have high electron affinities. For example, sodium (Na), which has a single electron in its outermost shell, can achieve a stable electronic configuration by losing its valence electron. This loss transforms sodium into a Na^+ ion. Conversely, chlorine (Cl), which requires one electron to complete its valence shell, can accept an electron, thereby becoming a Cl^- ion.

The formation of ions can be illustrated with the following chemical equation:
$$Na \rightarrow Na^+ + e^-$$
$$Cl + e^- \rightarrow Cl^-$$

Upon formation, these ions, Na^+ and Cl^-, attract each other because of their opposite charges, leading to the formation of an ionic bond. The overall reaction can be represented as:
$$Na + Cl \rightarrow Na^+ + Cl^- \rightarrow NaCl$$

Lattice Energy and Stability of Ionic Compounds

The strength of an ionic bond is significantly affected by lattice energy, which is the energy released when the gaseous ions form an ionic solid. Lattice energy is inversely proportional to the sum of the ionic radii; higher lattice energy corresponds to a smaller distance between ions and a stronger attraction.

This energy can be estimated using the Born-Haber cycle, a thermodynamic cycle that involves several steps to determine the overall energy change for the formation of an ionic compound from its constituent elements. For NaCl, the cycle involves the following steps:

1. Sublimation of solid sodium to gaseous atoms.
2. Ionization of sodium atoms to form cations.
3. Dissociation of chlorine molecules into chlorine atoms.
4. Addition of an electron to chlorine atoms to form anions.
5. Formation of the ionic compound from the gaseous ions.

Mathematically, the lattice energy U can also be estimated using the formula derived from Coulomb's Law:
$$U = \frac{k \cdot Q_1 \cdot Q_2}{r}$$

where k is a proportionality constant, Q_1 and Q_2 are the charges of the ions, and r is the distance between the centers of the ions.

Properties of Ionic Compounds

Ionic compounds exhibit a range of distinctive properties:

- High melting and boiling points due to the strong electrostatic forces between ions.
- Typically solid at room temperature.
- When dissolved in water or melted, ionic compounds are good conductors of electricity due to the mobility of ions.
- Generally soluble in polar solvents but insoluble in non-polar solvents.

To conclude, ionic bonding involves the complete transfer of one or more electrons from a metal to a non-metal. The resultant electrically charged ions exhibit strong electrostatic attractions, forming a crystalline lattice that gives ionic compounds their unique characteristics. Understanding these concepts is crucial for predicting the behavior of such compounds in various chemical and physical contexts.

4.3 Covalent Bonding: Sharing of Electrons

Covalent bonding represents a fundamental type of chemical bond where electrons are shared between atoms. This sharing allows each atom to attain a more stable electronic configuration, often approaching that of a noble gas. The concept of covalent bonding is crucial in explaining the structure and behavior of a vast array of molecular substances, ranging from simple diatomic molecules like hydrogen (H_2) to complex organic compounds and biochemical macromolecules.

Mechanism of Electron Sharing

In a covalent bond, the shared electrons contribute to the outer electron shells of both bonded atoms. Consider a simple molecule such as hydrogen chloride (HCl). The hydrogen atom ($1s^1$) and the chlorine atom ($3s^2 3p^5$) can achieve greater stability through sharing their valence electrons, forming a covalent bond.

$$H \cdot + \cdot Cl \rightarrow H : Cl$$

The electron dot formula, as shown above, illustrates that each atom contributes one unpaired electron, which pair up to form a bond. This shared pair of electrons is attracted to the nuclei of both atoms, which effectively lowers the potential energy of the system, resulting in a more stable molecular structure.

The Role of Electronegativity in Covalent Bonds

Electronegativity plays a significant role in determining the nature of the electron sharing in covalent bonds. It is a measure of the tendency of an atom to attract a bonding pair of electrons. In the HCl molecule, chlorine is more electronegative than hydrogen, hence, the shared pair of electrons is more attracted to the chlorine atom, creating a polar covalent bond. This polarity arises due to the uneven distribution of electron density between the involved atoms.

The mathematical expression for the dipole moment (μ) can be used to quantify the polarity of a bond:

$$\mu = \delta \times d$$

where δ represents the magnitude of the partial charges (in coulombs) on the atoms involved, and d signifies the distance (in meters) between these charges. The dipole moment is an important concept as it influences many physical properties such as boiling point, solubility, and reactivity.

Lewis Structures and the Octet Rule

The Lewis structure is a valuable tool in representing covalent bonds. It involves the use of dots to represent electrons and lines to symbolize covalent bonds. The octet rule, which states that atoms tend to bond in such a way that each atom has eight electrons in its valence

4.3. COVALENT BONDING: SHARING OF ELECTRONS

shell, guides the drawing of Lewis structures for most molecules. Exceptions to this rule include molecules like BF_3, where boron has only six valence electrons in the bonded state.

To illustrate this, let us consider the molecule of water (H_2O). Oxygen has six valence electrons and needs two more to complete its octet, which it acquires by forming two covalent bonds with two hydrogen atoms:

$$H — O — H$$

Each line between the atoms represents a pair of shared electrons, and the lone pairs on oxygen are shown as pairs of dots, emphasizing the completion of the octet rule.

Covalent Bond Strength and Bond Energy

The strength of a covalent bond is described by its bond dissociation energy, which is the energy required to break the bond between two covalently bonded atoms. Each type of bond has a characteristic bond energy; for instance, the average bond energy of a C-C single bond is approximately 347 kJ/mol, while that of a C=C double bond is about 614 kJ/mol.

Bond energies are intrinsically linked to molecular stability and reactivity. Molecules with higher bond dissociation energies are generally more stable and less reactive under normal conditions. This principle is fundamentally important in understanding the kinetics and mechanisms of chemical reactions.

Molecular Orbital Theory Integration

While discussing covalent bonding, it is imperative to consider molecular orbital theory, which provides a more detailed and quantitative picture of electron distribution in molecules. In molecular orbital theory, atomic orbitals of bonding atoms combine to form new orbitals called molecular orbitals. These can be bonding or antibonding, depending on their effect on the stability of the molecule.

The overlap of p orbitals in an oxygen molecule (O_2) results in the formation of both bonding (σ_p) and antibonding (σ_p^*) molecular orbitals, as represented below:

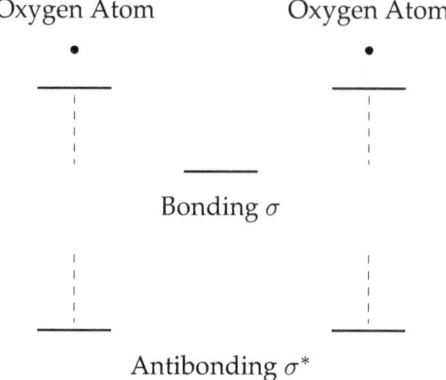

This integration of molecular orbital theory not only complements the Lewis structure approach but also enhances our understanding by providing insights into the electronic properties and stability of molecules from a quantum mechanical perspective.

Overall, covalent bonding remains central to the formation and function of molecules. Mastery of this concept is essential for understanding more complex chemical behaviors and the structural underpinnings of materials and life itself.

4.4 Metallic Bonding: Electron Sea Model

Metallic bonding is a type of chemical bond that forms between atoms of metallic elements. It involves the delocalization of valence electrons, resulting in a sea of electrons that surrounds a lattice of positive ions. This model, also known as the electron sea model, explains many physical properties of metals, including conductivity, malleability, and ductility.

In the electron sea model, atoms in a metal release some of their electrons from the outermost electron shell. These electrons are not bound to any specific atom but move freely within the metal lattice. The structure of a metallic bond can be visualized as a three-dimensional array of positive ions (cations) immersed in a cloud of delocalized, free-moving electrons.

This electron delocalization occurs because metals have relatively low ionization energies and low electronegativities, facilitating the

4.4. METALLIC BONDING: ELECTRON SEA MODEL

loss of electrons. The free electrons act as a glue holding the positive ions together, creating a strong bond between them. The nature of this bond is electrical and non-directional, meaning that the strength of bonding is equal in all directions around each ion. This isotropic bonding nature is what gives metals their unique ability to be hammered into thin sheets (malleability) and drawn into wires (ductility).

The mathematical description of bonding and behavior of electrons in metals can be modeled using quantum mechanical frameworks. One popular method is to apply the free electron model, wherein electrons are treated as free particles within a potential box defined by the dimensions of the metallic crystal. The solution of Schrödinger's equation for these electrons gives rise to a series of energy levels, or bands. The specific arrangement and occupancy of these electron bands are responsible for many properties of metals.

For instance, the electrical conductivity of metals can be explained through band theory. In conductors, the valence band containing the delocalized electrons partially overlaps with an empty conduction band. As a result, electrons can move freely under the influence of an electric field, facilitating the flow of electric current. Moreover, as electrons are delocalized, they can move to neighboring atoms with minimal resistance, a phenomenon fundamentally supporting the high thermal conductivity observed in metals.

The metallic bond strength varies among different metals and is influenced by the number of delocalized electrons and the charge and size of the metal ions. Smaller ions with higher charges typically have stronger metallic bonds due to the compact nature and greater electrostatic forces within the lattice.

A practical examination of the electron sea model can be seen in the alloying process of metals. When different metals are alloyed, the overall distribution of electrons in the electron sea can change, affecting the physical properties of the resulting alloy. For example, the addition of carbon to iron, forming steel, significantly increases its hardness and tensile strength. This is due to the influence of carbon atoms on the mobility of the delocalized electrons in the iron lattice, thereby affecting the distribution and density of the electron sea.

The implications of metallic bonding stretch also into chemical reactivity, corrosion resistance, and even catalytic behavior of metals. The availability and mobility of electrons can greatly influence how metals interact with their environment and other chemical substances.

The electron sea model, while simplified, provides essential insights into understanding the fundamental aspects of metallic properties. Further refined theories and computational models continue to build upon this basic understanding to delve deeper into the complex nature of metals and their interactions, essential for materials science and engineering applications.

4.5 Molecular Orbital Theory

Molecular Orbital Theory (MOT) represents a fundamental conceptual approach in quantum chemistry, providing an essential framework for understanding the electronic structure of molecules. According to this theory, atomic orbitals (AOs) of the constituent atoms combine to form molecular orbitals (MOs) when a molecule is formed. These MOs are spread over the entire molecule and can be occupied by electrons from any of the bonded atoms, contrasting the localized electron model seen in Lewis structures and valence bond theory.

Formation of Molecular Orbitals

The formation of molecular orbitals is typically described by the linear combination of atomic orbitals (LCAO). Mathematically, a molecular orbital ψ can be expressed as:

$$\psi = c_a \phi_a + c_b \phi_b$$

where ϕ_a and ϕ_b are the atomic orbitals from two different atoms and c_a, c_b are the corresponding coefficients determining the contribution of each atomic orbital to the molecular orbital.

When two atomic orbitals combine, they form two molecular orbitals: a bonding molecular orbital (BMO) and an antibonding molecular orbital (ABMO). Electrons in BMOs have lower energy compared to the original AOs and serve to stabilize the molecule, whereas electrons in ABMOs have higher energy and can destabilize it.

Bonding and Antibonding Orbitals

The difference between bonding and antibonding orbitals can be visualized through their electron density distributions. In bonding or-

4.5. MOLECULAR ORBITAL THEORY

bitals, the electron density is concentrated between the nuclei, leading to an attractive force that holds the nuclei together. Conversely, in antibonding orbitals, there is a node between the nuclei where electron density is zero, creating a repulsive force that can prevent bond formation if these orbitals are occupied.

For example, consider the hydrogen molecule (H_2). The atomic orbitals of each hydrogen atom, which are 1s orbitals, combine to form:

$$\psi_{bonding} = \frac{1}{\sqrt{2}}(\phi_{1s1} + \phi_{1s2})$$

$$\psi_{antibonding} = \frac{1}{\sqrt{2}}(\phi_{1s1} - \phi_{1s2})$$

where ϕ_{1s1} and ϕ_{1s2} are the 1s orbitals of the two hydrogen atoms.

Molecular Orbital Diagrams

Molecular orbital diagrams are valuable tools for visualizing the relative energies and occupancy of MOs. A typical MO diagram for H_2 includes energetically lower bonding MOs filled with electrons, followed by higher-energy antibonding MOs, which might stay unoccupied in lower energy states.

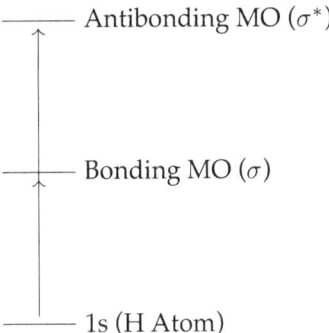

Applications of Molecular Orbital Theory

MOT is not restricted to the simplest molecules but extends efficiently to polyatomic and complex systems. It helps predict magnetic properties, molecular stability, and reactivity. For molecules with delocalized electrons, such as benzene, MOT provides a clearer and more accurate depiction of electron distribution than other theories.

MOT also plays a critical role in spectroscopy, where transitions between energy levels within molecular orbitals can be correlated with electromagnetic radiation absorption, providing profound insight into electronic structure.

Furthermore, MOT assists in understanding the bonding in metal complexes and coordination compounds by describing the interactions between metal d-orbitals and ligand orbitals, thus explaining the geometrical structure and reactivity of these compounds.

Molecular Orbital Theory enriches our understanding of molecular structures by describing how atomic orbitals combine to form molecular orbitals, defining the electronic configurations within molecules, and elucidating the intrinsic properties derived from these configurations. By elaborating on the formation, nature, and applications of molecular orbitals, MOT remains a cornerstone of modern chemical theory and practice.

4.6 Hybridization in Covalent Bonds

Hybridization is the concept in molecular orbital theory that describes the mixing of atomic orbitals to form new hybrid orbitals, which can then be used to form covalent bonds with other atoms. This phenomenon is pivotal in understanding how atoms combine to form molecules with specific shapes and properties.

The basic types of hybridization commonly encountered in organic and inorganic chemistry are sp, sp^2, and sp^3 hybridization. Each type of hybridization involves orbitals from the central atom mixing in specific ratios to achieve optimal overlapping with orbitals from other bonding atoms, thus forming more stable molecules.

sp^3 Hybridization

In sp^3 hybridization, one s orbital mixes with three p orbitals to form four equivalent sp^3 hybrid orbitals. Each of these orbitals has 25% s character and 75% p character. These hybrid orbitals are oriented tetrahedrally with an angle of 109.5° between them. Molecules such as methane (CH_4) exemplify sp^3 hybridization.

4.6. HYBRIDIZATION IN COVALENT BONDS

The molecular orbital formation can be described mathematically:

$$sp_1^3 = \frac{1}{2}(s + p_x + p_y + p_z)$$
$$sp_2^3 = \frac{1}{2}(s + p_x - p_y - p_z)$$
$$sp_3^3 = \frac{1}{2}(s - p_x + p_y - p_z)$$
$$sp_4^3 = \frac{1}{2}(s - p_x - p_y + p_z)$$

sp^2 Hybridization

sp^2 hybridization occurs when one s orbital and two p orbitals hybridize to form three sp^2 hybrid orbitals, each with 33% s character and 67% p character. The unhybridized p orbital remains perpendicular to the plane of the three hybridized orbitals. This configuration leads to trigonal planar geometry with bond angles of 120°. A typical molecule exhibiting sp^2 hybridization is ethylene (C_2H_4).

The hybrid orbitals can be expressed as:

$$sp_1^2 = \frac{1}{\sqrt{3}}(s + p_x + p_y)$$
$$sp_2^2 = \frac{1}{\sqrt{3}}(s + p_x - p_y)$$
$$sp_3^2 = \frac{1}{\sqrt{3}}(s - 2p_x)$$

sp Hybridization

In sp hybridization, one s orbital mixes with one p orbital, resulting in two sp hybrid orbitals that are oriented linearly, 180 degrees apart. This type of hybridization is characterized by having 50% s character and 50% p character. Acetylene (C_2H_2) is an example, where the carbon atoms are bonded via a triple bond consisting of one σ bond and two π bonds.

The mathematical form for sp hybrid orbitals can be described as:

$$sp_1 = \frac{1}{\sqrt{2}}(s + p_x)$$
$$sp_2 = \frac{1}{\sqrt{2}}(s - p_x)$$

Through these expressions and examples, it can be seen how hybridization underlies the structure and bonding in molecules. Understanding the configuration and characteristics of hybrid orbitals helps in predicting the geometry and reactivity of molecules, making it an essential concept in molecular chemistry.

Further, it is the geometry resulting from these hybridizations that markedly affect the physical and chemical properties of compounds. Given the arrangement of the orbits, molecules can exhibit different angles and spatial distributions, which in turn influence molecular interactions, boiling points, melting points, and their overall stability.

To validate these theoretical models, spectroscopic studies like infrared spectroscopy and X-ray crystallography can be employed. These methods provide empirical data that confirm the predicted structures and the theoretical calculations of hybrid orbitals, supporting the fundamental principles of quantum chemistry covered in this chapter.

4.7 Valence Shell Electron Pair Repulsion (VSEPR) Theory

The Valence Shell Electron Pair Repulsion (VSEPR) theory provides a systematic approach to understanding the arrangement of atoms around a central atom in a molecule, based predominantly on minimizing the electrostatic repulsion between valence electron pairs in the valence shell of the central atom. Central to the application of VSEPR theory is the concept of molecular geometry, which fundamentally influences physical and chemical properties of a molecule.

According to VSEPR theory, electron pairs located in the valence shell of an atom will arrange themselves as far apart as possible to minimize repulsion. This behavior governs the three-dimensional geometry of the molecules. Electron pairs can be either bonding pairs,

4.7. VALENCE SHELL ELECTRON PAIR REPULSION (VSEPR) THEORY

shared between atoms, or lone pairs, which are unshared electrons localized on a single atom.

The steps in predicting the molecular geometry using VSEPR theory involve: 1. Determining the Lewis structure of the molecule. 2. Counting the total number of electron pairs (bonding and lone pairs) around the central atom. 3. Determining the arrangement of these electron pairs based on minimizing repulsions, which leads to specific molecular geometries.

The common molecular geometries obtained through VSEPR theory include:

- Linear
- Trigonal planar
- Tetrahedral
- Trigonal bipyramidal
- Octahedral

Each geometry corresponds to a specific number of electron pairs around the central atom:

Electron Pairs	Geometry	Example
2	Linear	CO_2
3	Trigonal planar	BF_3
4	Tetrahedral	CH_4
5	Trigonal bipyramidal	PCl_5
6	Octahedral	SF_6

Table 4.1: VSEPR geometries and examples

An important aspect to consider is that lone pairs exert a greater repulsive force on adjacent electron pairs than bonding pairs. This is primarily due to lone pairs being closer to the nucleus and thus more localized in space. Therefore, when predicting molecular shapes, adjustments must be made for the enhanced repulsion from lone pairs.

For example, in the case of water (H_2O), which has a tetrahedral arrangement of electron pairs, two are bonding pairs and two are lone pairs. The repulsion between the lone pairs forces the bonding pairs closer together, resulting in a bent molecular geometry instead of a flat tetrahedral shape.

To illustrate the application of VSEPR theory, let us consider ammonia (NH_3):

1. Ammonia has one lone pair and three bonding pairs around the nitrogen atom.

2. These four pairs of electrons would initially suggest a tetrahedral geometry.

3. However, the lone pair-bonding pair repulsions adjust the geometry, pushing the hydrogen atoms closer, resulting in a trigonal pyramidal shape.

The bond angles in NH_3 are approximately 107 degrees, slightly less than the 109.5 degrees expected in a perfect tetrahedral geometry due to the repulsion from the lone pair.

In sum, VSEPR theory is a critical tool in the chemist's arsenal for predicting the geometries of molecules. Its application, while straightforward in many cases, may require careful consideration of the effects of lone pairs and multiple central atoms in larger and more complex molecules.

4.8 Molecular Geometry and Polarity

Molecular geometry and polarity are vital concepts that determine the physical and chemical properties of molecules. The arrangement of atoms within a molecule and the distribution of electrons dictate the shape and polarity of the molecule, affecting interactions such as hydrogen bonding, solubility, boiling point, and reactivity.

Determining Molecular Geometry

The molecular geometry of a compound can be predicted using the Valence Shell Electron Pair Repulsion (VSEPR) theory, which was introduced in previous sections. According to VSEPR theory, the shape of a molecule is determined by the repulsions between electron pairs in the valence shell of the central atom. The theory can be summarized with the following steps:

1. Count the total number of valence electrons for the central atom.

4.8. MOLECULAR GEOMETRY AND POLARITY

2. Add electrons based on the negative charge or subtract based on the positive charge of the molecule.

3. Determine the total number of electron pairs (bonding pairs and lone pairs) around the central atom.

4. Apply the VSEPR model to predict the molecular shape based on minimizing repulsions between electron pairs.

Common Molecular Geometries

The geometry is often described by the basic shape that considers electron domains (bonding and non-bonding electron pairs). Common molecular geometries include linear, trigonal planar, tetrahedral, trigonal bipyramidal, and octahedral, described below with bond angles and examples:

1. Linear geometry (180° bond angle) - e.g., CO_2, HCN

2. Trigonal planar (120° bond angle) - e.g., BH_3, $AlCl_3$

3. Tetrahedral (109.5° bond angle) - e.g., CH_4, NH_4^+

4. Trigonal bipyramidal (90° and 120° bond angles) - e.g., PCl_5, $Fe(CO)_5$

5. Octahedral (90° bond angle) - e.g., SF_6, WCl_6

Special consideration is necessary for cases involving lone pairs as these pairs repel more strongly than bonding pairs, often leading to distortion in ideal bond angles.

Molecular Polarity

Molecular polarity depends on both the molecular geometry and the electronegativity of the atoms involved. It can be determined through the following steps:

1. Identify polar bonds within the molecule by comparing the electronegativity of bonded atoms.

2. Assess the symmetry of the molecule. Symmetrical molecules with polar bonds may still be overall nonpolar if the bond polarities cancel each other.

3. Use vector addition for dipole moments: if the vector sum of the polar bonds (dipoles) does not cancel out, the molecule is polar.

A non-zero dipole moment indicates a polar molecule, leading to asymmetric distribution of electron density. Polarity impacts many physical properties, including solubility, melting and boiling points, and reactivity.

Example of Determining Polarity

Consider water molecule (H_2O). Oxygen is more electronegative than hydrogen, creating two polar bonds. Water's bent molecular geometry (with a bond angle of approximately 104.5°) leads to an asymmetric distribution of electron density, thus a net dipole moment and a polar molecule.

Illustration with TikZ

The following TikZ example provides a visual representation of the molecular geometry of H_2O:

This diagram not only aids in visual learning but also emphasizes the concept of vector addition of dipoles in determining overall polarity.

: The detailed understanding of molecular geometry and polarity crucially enhances the prediction of molecular behavior in different chemical environments, aiding in the rational design of molecules for specific functions in materials science, pharmacology, and other fields of chemistry.

This segment integrates principles of molecular geometry and polarity from initial determination steps through to visualization, offering a thorough comprehension suitable for learners at different levels.

4.9 Intermolecular Forces

Integrating within our discourse on molecular structures and chemical bonding, this section focuses on intermolecular forces (IMFs), pivotal in dictating the physical and some chemical properties of substances. These forces are essentially the attractions or repulsions that act between molecules or between non-bonded atoms. The understanding of IMFs aids in comprehending boiling and melting points, solubility, and various states of matter.

Types of Intermolecular Forces

There are primarily three types of intermolecular forces:

1. **London Dispersion Forces:** Also known as dispersion forces or induced dipole-induced dipole interactions, these are the weakest type of IMF. They arise due to the momentary uneven distribution of electrons around an atom or molecule, creating a temporary dipole. London dispersion forces are universal and occur between all atoms and molecules, increasing in strength with the number of electrons.

2. **Dipole-Dipole Interactions:** These forces appear between two polar molecules. The partial positive charge of one molecule is attracted to the partial negative charge of the neighboring molecule. Their strength depends on the polarity of the molecules.

3. **Hydrogen Bonding:** Though often considered a separate category due to its relative strength, hydrogen bonding is a particularly strong form of dipole-dipole interaction. It occurs when hydrogen is bonded to a highly electronegative element (usually nitrogen, oxygen, or fluorine), allowing for significant partial charges.

Quantitative Analysis of Intermolecular Forces

The strength of IMFs can be quantitatively described by considering both molecular properties and environmental conditions. We calculate the potential energy (U) associated with the interactions through the general expression for electrostatic potential energy:

$$U = \frac{kQ_1Q_2}{r}$$

where U denotes the potential energy, k is Coulomb's constant, Q_1 and Q_2 are the magnitudes of the charges, and r is the distance between the charges. For dipole-dipole interactions, the equation becomes:

$$U = \frac{kQ\mu}{r^2}$$

where μ represents the dipole moment of the molecule.

Applications and Implications of Intermolecular Forces

1. **Melting and Boiling Points:** Substances with strong IMFs require higher temperatures to overcome these forces, hence possess higher boiling and melting points. 2. **Solubility:** The solubility of substances in different solvents can often be predicted based on IMFs. Generally, 'like dissolves like' — polar compounds are more likely to dissolve in polar solvents, and non-polar compounds in non-polar solvents. 3. **Viscosity:** The resistance of a liquid to flow, or its viscosity, increases with stronger IMFs because more energy is needed to allow molecules to move past each other. 4. **Surface Tension:** This is a measure of the energy required to increase the surface area of a liquid due to intermolecular forces. High surface tension in water, largely due to hydrogen bonding, allows for phenomena such as water striding insects.

To analyze these phenomena in depth, consider a practical example wherein acetone (C_3H_6O) and water (H_2O) are mixed. Both substances are polar and capable of hydrogen bonding, yet their interaction strengths and resulting properties differ markedly due to differences in molecular structure and hydrogen bonding capabilities. This differential analysis assists in understanding why acetone is a good solvent for many organic compounds but not for others, like fats.

Intermolecular forces are critical not only to chemical bonding and structure but also fundamentally influence the everyday physical properties of materials. Their extensive implications make them a crucial topic for study in chemical sciences.

4.10 Resonance Structures in Molecular Stability

Resonance is a fundamental concept in the study of molecular structure and stability within the context of quantum chemistry. It describes a scenario in which two or more Lewis structures with the same arrangement of atoms but different distributions of electrons can be drawn for a molecule or ion. These alternative structures, known as resonance structures, are not real and distinct entities; rather, they are hypothetical constructs that provide a clearer understanding of the actual, intermediate structure of the molecule, which often exhibits enhanced stability due to this electron delocalization.

The Concept of Resonance

Each Lewis structure, drawn to represent a resonating molecule, is termed a resonance form or contributor. It is essential to note that these forms are purely fictional; the true molecular structure is a hybrid (or a resonance hybrid) of these forms and not any individual form. The resonance hybrid is represented by a structure reflecting the average positions of all electrons as suggested by the contributing forms, often delineated using partial bonds or fractional charges.

Electron Delocalization and Stability

In resonance structures, the delocalization of pi (π) electrons across multiple atoms in a molecule leads to a lowering of the overall potential energy, thereby increasing the stability of the molecule. Mathematically, this can be expressed by considering the resonance energy, which is the difference in energy between the resonance hybrid and the most stable contributing resonance form. A greater resonance energy generally correlates with greater molecular stability.

For example, consider benzene (C_6H_6), a key example in resonance study due to its aromatic nature. Benzene can be described by two resonance structures, each showing alternating double and single bonds forming a hexagonal ring. Neither of these structures alone can account for benzene's known chemical behavior and properties. Instead, we understand that the actual molecular structure of benzene is a resonance hybrid of these two forms, with delocalized pi electrons spread evenly across all six carbon atoms:

Such electron delocalization not only achieves a lower potential energy but also results in increased symmetry and equivalency among bonds, manifest in all carbon-carbon bonds in benzene exhibiting equal length, intermediate between that of a single and a double bond.

Rules for Writing Valid Resonance Structures

To effectively use resonance in the analysis of molecular stability, various key guidelines should be followed in drawing valid resonance structures:

- All resonance forms must contain the same number of electrons and adhere to the conservation of charge.

- Atoms can never be moved; only electrons can be shifted in the form of pi electrons or lone electron pairs.

- Each contributing structure must comply with the rules of normal valency for the atoms involved.

- Contributions of various resonance structures to the hybrid are not necessarily equal; structures with lower energies and fewer charge separations generally contribute more to the hybrid.

Quantifying Resonance with Computational Tools

Advances in computational chemistry allow for the quantification of resonance and its effects on molecular stability. Techniques such as molecular orbital theory (discussed earlier in this chapter) and computational methods including Hartree-Fock and density functional theory provide detailed insights into the electron density distribution and energy states in molecules, enabling a deeper understanding of resonance effects at an electronic level.

These computational models, by allowing for the visualization and measurement of electron delocalization, enhance the comprehensibility and predictability of resonance effects on molecular stability—crucial for the design and synthesis of new molecular structures in pharmaceuticals, materials science, and other applied fields.

By exploring the foundations and implications of resonance, discerning learners and chemists alike gain valuable insights into not only molecular stability but the electronic behavior underlying fundamental chemical properties.

4.11 Hydrogen Bonding and its Special Properties

Hydrogen bonding represents a special type of dipole-dipole interaction that occurs between a hydrogen atom, which is covalently bonded to a highly electronegative atom (typically oxygen, nitrogen, or fluorine), and an electronegative atom with a lone pair of electrons. This section elaborates on the properties of hydrogen bonds and their implications on the physical and chemical characteristics of compounds.

When a hydrogen atom forms a covalent bond with an electronegative atom like oxygen or nitrogen, the unequal sharing of electrons results in a partial positive charge (δ^+) on hydrogen and a partial negative charge (δ^-) on the more electronegative atom. This δ^+ on hydrogen is attracted to the δ^- on the lone pair of an adjacent electronegative atom, forming a hydrogen bond, denoted as

$$A - H \cdots B$$

where A and B are electronegative atoms, and H is hydrogen.

The strength of hydrogen bonds is typically in the range of 5 to 30 kJ/mol, considerably weaker than covalent bonds but stronger than most other types of van der Waals interactions. This moderate strength influences various properties, such as boiling point, viscosity, and solubility.

- **Boiling point:** Substances with hydrogen bonds generally exhibit higher boiling points. For instance, water (H_2O), with extensive hydrogen bonding, has a much higher boiling point than expected for a molecule of its size.

- **Viscosity:** Hydrogen bonding can significantly increase the viscosity of a liquid. The movement of molecules in a liquid involves breaking and reforming of hydrogen bonds, which requires additional energy, thereby increasing resistance to flow.

- **Solubility:** Polar substances with potential for hydrogen bond formation are more likely to be soluble in solvents that can also form hydrogen bonds, due to the favorable interaction between similar types of molecules.

One special characteristic often associated with hydrogen bonding is its effect on the structure and properties of water. The extensive hydrogen bonding in water leads to properties like high surface tension, anomalously high boiling and freezing points, and the expansion of water upon freezing.

Hydrogen bonds also play a critical role in biochemical structures and processes. The double helical structure of DNA is stabilized by hydrogen bonds between the complementary base pairs. In proteins, hydrogen bonds are essential for the formation of secondary structures such as α-helices and β-sheets. These interactions are crucial for maintaining the proper 3D structure required for biological function.

In addition to naturally occurring scenarios, hydrogen bonding is exploited in various applications, including the design of polymers and the formulation of pharmaceuticals. For example, the tailored interaction of drug molecules with biological targets through hydrogen bonds can dramatically enhance the efficacy of a drug.

The properties of hydrogen bonds can be visualized using molecular modeling tools and spectroscopy techniques like Nuclear Magnetic Resonance (NMR) and Infrared Spectroscopy (IR). These techniques provide insights into the geometry and environment of hydrogen bonds in molecular systems.

4.12 Computational Predictions of Molecular Structures

Computational chemistry stands as a robust framework for analyzing and predicting molecular structures and behaviors, using the principles of quantum mechanics. The primary objective of computa-

4.12. COMPUTATIONAL PREDICTIONS OF MOLECULAR STRUCTURES

tional techniques in the prediction of molecular structures is to solve the Schrödinger equation for molecules, thereby determining energy states and spatial configurations of electrons which directly influence molecular geometry.

To begin, the fundamental field of computational molecular predictions is defined by two main approaches: *ab initio* methods and semi-empirical methods. *Ab initio* methods, which mean "from first principles", do not rely on experimental data but solely on theoretical principles. They solve quantum mechanical equations to calculate molecular properties purely from the charge, mass, and spin of the constituent nuclei and electrons. Common *ab initio* methods include Hartree-Fock (HF), Post-Hartree-Fock, and Density Functional Theory (DFT).

Hartree-Fock Method: The HF method approximates the wave function of a multielectron system as a single Slater determinant of one-electron wave functions called orbitals. Each electron is described as moving in an average field created by all other electrons. Mathematically, the HF method uses the Fock matrix to simplify the many-body problem into a series of single-body problems. The equations are solved iteratively until a self-consistent field (SCF) is achieved. The limitation here is the lack of correlation energy, which is the energy obtained from the specific interactions between pairs of electrons that are omitted in the average-field approximation.

Density Functional Theory (DFT): This method improves upon the HF approach by considering the electron density, a function of position, as the fundamental property, not the wave function. This shift significantly simplifies computations because the electron density is a function of three spatial coordinates, whereas the wave function encompasses $3N$ coordinates, N being the number of electrons. The exchange-correlation functional in DFT approximates the complexities of electron interaction, balancing accuracy and computational feasibility.

Semi-empirical methods, on contrast, involve some approximations and parameters derived from experimental data. Techniques such as the Austin Model 1 (AM1) and Parametric Method 3 (PM3) are computationally less intensive and are suitable for large molecules where the *ab initio* computation becomes infeasible.

To illustrate the computational prediction of molecular structures, consider benzene (C_6H_6), a common subject of study due to its aromatic properties. Using DFT, one can predict not only the planar geometry of the molecule, typically assumed due to the π bonding in the ring, but also its electronic properties like the distribution of π electrons above and below the ring plane.

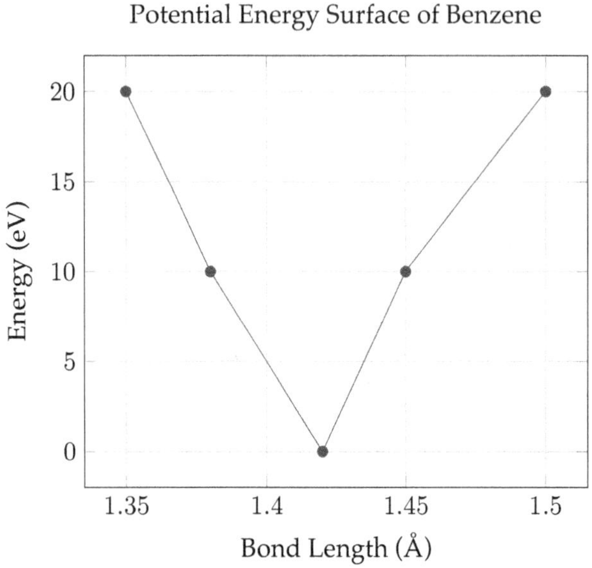

Potential Energy Surface of Benzene

Figure 4.1: Hypothetical plot demonstrating the potential energy surface of benzene as a function of carbon-carbon bond length

The landscape of molecular modeling also includes molecular mechanics, a method that uses classical physics to model molecular systems. The molecular mechanics apply potential energy functions to correlate the structure and properties. While quantum mechanical methods consider electrons explicitly, molecular mechanics treat atoms and bonds as balls and springs, defining potential energy in terms of bond stretching, angle bending, and torsional rotation, which is particularly useful in modeling biological macro-molecules like proteins.

Each of these computational techniques has its own suitability and is often chosen based on the size of the molecule, the properties of interest, computational resources, and required accuracy. In education, the selection of a particular method may also depend on the area of application such as medicinal chemistry, materials science, or theoretical chemistry.

Chapter 5

Quantum Mechanical Models and Approximations

This chapter focuses on the theoretical models and approximations used in quantum chemistry to solve complex systems that are otherwise analytically intractable. It begins with an introduction to the Born-Oppenheimer approximation, vital for simplifying molecular wavefunctions, and progresses through perturbation theory and the variational principle, which are essential for understanding molecular behavior. Discussions include the Hartree-Fock method and Density Functional Theory (DFT), both critical for computational quantum chemistry. By providing insights into these models, the chapter prepares readers to tackle real-world chemical problems using quantum mechanical tools.

5.1 Overview of Quantum Mechanical Models

Quantum mechanical models are pivotal in understanding the electronic structure of molecules and the behavior of atoms within a molecule. The underlying theory of quantum mechanics provides a framework by which molecular structure, reactivity, physical properties, and spectroscopy can be accurately studied. The basis of all quantum mechanical models rests on the fundamentals of wavefunctions, operators, and quantum postulates.

A wavefunction, denoted as ψ, is a mathematical function that describes the quantum state of a system. The probability density of the position of electrons in a molecule is given by $|\psi|^2$. The entire behavior of a system in quantum mechanics is determined by the Schrödinger equation, an essential component in the study of chemical systems:

$$\hat{H}\psi = E\psi$$

where $\hat{H}$ is the Hamiltonian operator representing the total energy of the system, and E is the energy eigenvalue associated with the wavefunction ψ.

Operators in quantum mechanics are mathematical objects that act on the wavefunctions and correspond to observable physical properties. The position and momentum are typical examples of such observables. Each observable in quantum mechanics is described by an operator, whose eigenvalues correspond to the measurement values.

Understanding the mathematical form and physical significance of Hamiltonian in quantum chemistry is crucial. For most systems of interest, $\hat{H}$ includes terms that account for kinetic energy, potential energy, and electron interactions:

$$\hat{H} = -\frac{\hbar^2}{2m}\nabla^2 + V$$

where ∇^2 is the Laplacian that represents the kinetic energy operator of the electrons, m is the mass of an electron, $\hbar$ is the reduced Planck's constant, and V is the potential energy, which includes electron-nuclear and electron-electron interactions.

This leads us to the implementation of various models and approximations to solve the Schrödinger equation for molecular systems. The exact solution is only feasible for the simplest systems like hydrogen atom. For more complex systems, we rely on approximations:

1. **The Born-Oppenheimer Approximation** simplifies molecular wavefunctions and energy calculations by assuming that the nuclear motion and electron dynamics can be separated owing to the large difference in their masses.

2. **Perturbation Theory** provides a method to calculate corrections to the first approximation results when the system Hamiltonian can

be divided into a solvable zeroth-order part plus a perturbative part.

3. **Variational Principle** is a method that states that the ground state of a Hamiltonian is the lowest possible expectation value of that Hamiltonian, obtained from all possible normalized wavefunctions. This principle leads to the formulation of variational methods used extensively for approximate solutions.

4. **Hartree-Fock Method** and **Density Functional Theory (DFT)** are computational methods used to derive approximations to the wavefunction and the corresponding energy of the electrons in molecular systems. They each use a different approach to treat electron correlation and computational complexity, greatly expanding the types of systems that can be studied quantum mechanically.

5. **Semi-empirical and Empirical methods** reduce the computational cost by exploiting approximations and parameterizations from experimental data enabling quick and reasonably accurate predictions of molecular properties.

6. Simulation methods such as **Monte Carlo Simulations and Molecular Dynamics** allow the exploration of molecular systems under various conditions by statistically sampling and using classical mechanics principles, respectively. These simulations are often integrated with quantum mechanical calculations to provide insights into the properties of large biomolecular systems over extended periods.

Thus, the various models and approximations in quantum mechanics each contribute uniquely to the comprehensive understanding and prediction of chemical phenomena, facilitating a deeper connection between theoretical calculations and experimental results.

5.2 The Born-Oppenheimer Approximation

The Born-Oppenheimer approximation, developed by physicists Max Born and J. Robert Oppenheimer, is a cornerstone in quantum mechanics, especially in the study of molecular systems. This approximation simplifies the complex quantum equations by leveraging the vast difference in mass between electrons and nuclei within a molecule. Notably, electrons are approximately 1836 times lighter than protons, which allows them to respond much more rapidly to changes in the positions of nuclei.

Principle and Application

In the Born-Oppenheimer approximation, the total molecular wavefunction $\Psi(\mathbf{r}, \mathbf{R})$, which encompasses both electronic coordinates $\mathbf{r}$ and nuclear coordinates $\mathbf{R}$, is simplified by treating the nuclear and electronic motions separately. This results in two distinct sets of equations: one for the electronic structure, considering the nuclei as fixed points ($\mathbf{R}$ treated as parameters), and one for the nuclear motion, which is influenced by the potential energy surfaces (PES) derived from the electronic structure calculations.

Mathematically, this separation is expressed as:

$$\Psi(\mathbf{r}, \mathbf{R}) \approx \chi(\mathbf{R})\psi(\mathbf{r}; \mathbf{R})$$

where $\psi(\mathbf{r}; \mathbf{R})$ represents the electronic wavefunction, dependent on electronic coordinates $\mathbf{r}$ and parametrically on the nuclear coordinates $\mathbf{R}$; $\chi(\mathbf{R})$ is the nuclear wavefunction.

Solving the Electronic Schrödinger Equation

The first step under this approximation involves solving the electronic Schrödinger equation:

$$\hat{H}_e \psi(\mathbf{r}; \mathbf{R}) = E_e(\mathbf{R})\psi(\mathbf{r}; \mathbf{R})$$

Here, $\hat{H}_e$ denotes the electronic Hamiltonian, which includes the kinetic energy of electrons and their interaction potentials with the stationary nuclei at coordinates $\mathbf{R}$. $E_e(\mathbf{R})$ is the electronic energy, which acts as a potential energy landscape in solving the nuclear Schrödinger equation.

Nuclear Schrödinger Equation

Given the electronic potential energy surface $E_e(\mathbf{R})$, the nuclear Schrödinger equation can be formulated as:

$$\left[-\frac{\hbar^2}{2M} \nabla_{\mathbf{R}}^2 + E_e(\mathbf{R}) \right] \chi(\mathbf{R}) = E\chi(\mathbf{R})$$

where $\nabla_{\mathbf{R}}^2$ is the Laplacian operator with respect to the nuclear coordinates, M denotes the mass of the nuclei, and E is the total energy of the system. This equation determines the vibrational and rotational states of the molecule.

Although the Born-Oppenheimer approximation simplifies many complex quantum mechanical calculations, it introduces the Born-Oppenheimer diagonal correction, which can be significant in systems involving light atoms or strongly coupled electronic and nuclear motions. Despite these limitations, the efficiency and insights provided by this approximation are indispensable, especially in computational chemistry where a full quantum mechanical treatment of all particles in a sizable molecule is beyond current computational means. It remains a fundamental tool in the theoretical and practical application of molecular quantum mechanics.

5.3 Perturbation Theory: Concept and Applications

Perturbation theory is an indispensable mathematical tool employed in quantum chemistry to find an approximate solution to a complex problem by building on the exact solution of a simpler, related problem. This approach is particularly useful when the system of interest can be seen as a small modification or "perturbation" of a system whose exact solution is known.

Basics of Perturbation Theory

In quantum mechanics, the fundamental problem is to solve the Schrödinger equation for the system:

$$\hat{H}\psi = E\psi,$$

where $\hat{H}$ is the Hamiltonian operator, ψ is the wavefunction of the system, and E is the energy eigenvalue corresponding to ψ. In cases where $\hat{H}$ is too complex due to interactions within the system, we decompose it as:

$$\hat{H} = \hat{H}_0 + \hat{V},$$

where $\hat{H}_0$ is the Hamiltonian of a solvable system and $\hat{V}$ is a small perturbative addition.

The foundational assumption in perturbation theory is that the energies and wavefunctions of $\hat{H}$ can be expanded in a power series around the known solutions of $\hat{H}_0$. The wavefunction and energy expansions are given by:

$$\psi = \psi_0 + \lambda\psi_1 + \lambda^2\psi_2 + \ldots,$$

$$E = E_0 + \lambda E_1 + \lambda^2 E_2 + \ldots,$$

where λ is a small parameter quantifying the strength of the perturbation, ψ_i and E_i for $i \geq 1$ are the corrections at the i-th order, with ψ_0 and E_0 being the zeroth-order terms (solutions from $\hat{H}_0$).

Application of Perturbation Theory

Let us consider the first-order correction terms (λ^1) which are most commonly used due to their simplicity and considerable accuracy in many cases. The first-order energy correction is computed as:

$$E_1 = \langle\psi_0|\hat{V}|\psi_0\rangle,$$

where $\langle\cdot|\cdot\rangle$ denotes the scalar product in Hilbert space.

The first-order correction to the wavefunction, ψ_1, is obtained by solving:

$$(\hat{H}_0 - E_0)\psi_1 = (\hat{V} - E_1)\psi_0,$$

where care must be taken to ensure that ψ_1 is orthogonal to ψ_0.

Examples and Implementation

To illustrate, consider a hydrogen atom subject to an external electric field as a perturbation. Here, $\hat{H}_0$ is the Hamiltonian of the hydrogen atom without the field, and $\hat{V}$ might represent the interaction of the electric dipole moment of the atom with the field. Calculations then follow to determine how the energy levels shift as a function of the field strength, reflecting phenomena such as the Stark effect.

Perturbation theory also finds extensive application in the study of molecular vibrations, where small displacements from equilibrium positions are treated as perturbations.

Limitations

Despite its utility, perturbation theory holds its accuracy only for sufficiently small values of λ. For large perturbations, higher-order terms become significant, and the series convergence needs to be carefully analyzed. Additionally, perturbation theory might fail for degenerate states where $\hat{H}_0$ has eigenvalues that are closely spaced or equal, requiring special treatment termed "degenerate perturbation theory."

In summary, perturbation theory provides a crucial approximate method for addressing the complexity of quantum mechanical systems. It helps predict how small changes in a system affect its dynamics, facilitating insights into the system's stability and shifts in its properties under various influences.

5.4 Variational Principle and Method

The Variational Principle is a cornerstone in quantum mechanics, providing a powerful approach to approximating the ground state energy of a quantum system. This principle states that for any normalized trial wave function Ψ, the expectation value of the Hamiltonian $\hat{H}$, defined as $E[\Psi] = \langle \Psi | \hat{H} | \Psi \rangle$, will always be greater than or equal to the ground state energy E_0 of the Hamiltonian. Mathematically, this can be represented as:

$$E[\Psi] \geq E_0$$

where $E[\Psi]$ is often termed as the variational energy. The equality holds true if and only if Ψ is exactly the ground state wave function Ψ_0 of the Hamiltonian.

This principle forms the basis for the variational method, which is used to approximate the ground state of a given Hamiltonian. The method involves proposing a trial wave function $\Psi(\vec{r}; \alpha)$ that depends on one or more parameters α. The objective is to adjust these parameters to minimize the expectation value $E[\Psi]$.

$$E(\alpha) = \frac{\langle \Psi(\vec{r}; \alpha) | \hat{H} | \Psi(\vec{r}; \alpha) \rangle}{\langle \Psi(\vec{r}; \alpha) | \Psi(\vec{r}; \alpha) \rangle}$$

The optimization of α to find the minimum of $E(\alpha)$, denoted as $E_{\min}$, approximates the ground state energy E_0 as closely as the flexibility of the trial wave function allows.

Choosing the Trial Wave Function Choosing an appropriate trial wave function is crucial in the variational method. The choice strongly depends on the system under study and the balance required between computational simplicity and accuracy. Common choices include:

- Linear combinations of atomic orbitals (LCAO) for molecular systems.

- Slater determinants comprising single-particle orbitals for many-electron systems.

- Parametric functions such as Gaussian type orbitals (GTOs).

Computing the Expectation Value To compute $E(\alpha)$ efficiently, one needs to evaluate the integrals involving the kinetic energy and potential energy operators of the Hamiltonian. These integrals may often become complex depending on the choice of $\hat{H}$ and $\Psi(\vec{r}; \alpha)$, generally requiring numerical methods such as Monte Carlo integration or grid-based numerical integration for their evaluation.

Minimizing the Energy Functional The minimization of $E(\alpha)$ can be performed using various numerical optimization techniques like gradient descent, Newton-Raphson method, or more sophisticated methods like the conjugate gradient or quasi-Newton methods. The goal is to find the set of parameters α where the derivative of $E(\alpha)$ with respect to α is zero, indicating a potential minimum.

Example To illustrate the variational principle, consider a particle in a one-dimensional box with a trial wave function $\Psi(x) = x(1-x)e^{\alpha x}$. Here, the parameter to be varied is α. The objective is to compute $E(\alpha)$ and find α that minimizes this energy.

Applications The variational method is not only applicable to finding ground state energies but can also be extended to estimate excited states' energies by using orthogonal constraints with respect to the lower energy states. It is a pivotal method used across various fields like condensed matter physics, nuclear physics, and quantum chemistry, particularly in the study of large atomic and molecular systems where exact solutions are not feasible.

In summary, the variational method serves as a fundamental tool in quantum mechanics enabling the estimation of the lowest energy states of quantum systems within the framework provided by the versatility and choice of the trial wave functions. As computational capabilities grow, so too does the accuracy and applicability of the variational method in tackling complex systems in quantum chemistry.

This covers the key principles, methodologies, and applies the section on the Variational Principle and method, explaining its purpose, execution, and significance in quantum chemistry.

5.5 Hartree-Fock Theory and Self-Consistent Field

Hartree-Fock (HF) theory is a cornerstone of computational quantum chemistry, offering a simplified but effective approach to solving the Schrödinger equation for many-electron systems. The primary aim of HF theory is to determine the wavefunction of a multi-electron system that best approximates the ground state by representing the complex problem of many interacting electrons through a set of single-electron functions known as molecular orbitals.

In the framework of HF theory, the total electronic wavefunction of the system, Ψ, is approximated by a single Slater determinant composed of these molecular orbitals. This wavefunction thus respects the antisymmetric property required by the Pauli exclusion principle. Each molecular orbital, $\phi_i(r)$, is expressed as a linear combination of a predefined set of basis functions, usually atomic orbitals. This representation forms the foundation of the mathematical structure of HF theory.

To formulate these concepts mathematically, we consider N electrons in a molecule with the Hamiltonian given by

$$\hat{H} = -\sum_{i=1}^{N} \frac{\hbar^2}{2m} \nabla_i^2 - \sum_{A,i} \frac{Z_A e^2}{4\pi\epsilon_0 r_{iA}} + \sum_{i<j} \frac{e^2}{4\pi\epsilon_0 r_{ij}},$$

where r_{iA} is the distance between electron i and nucleus A, r_{ij} is the distance between electrons i and j, Z_A is the charge of nucleus A, e is the elementary charge, ϵ_0 is the permittivity of free space, and $\hbar$ is the reduced Planck constant.

The single-electron functions must satisfy the Fock equations, which are derived from the variational principle seeking to minimize the total electronic energy of the system. The Fock operator, $\hat{F}$, is defined as

$$\hat{F}_i = -\frac{\hbar^2}{2m}\nabla_i^2 - \sum_A \frac{Z_A e^2}{4\pi\epsilon_0 r_{iA}} + \hat{J}_i - \hat{K}_i,$$

where $\hat{J}_i$ and $\hat{K}_i$ represent the Coulomb and exchange operators, respectively. These operators account for electron-electron repulsions and the antisymmetry requirement, and they are defined by

$$\hat{J}_i \phi_i(r) = \sum_{j \neq i} \int \frac{e^2}{4\pi\epsilon_0 r_{ij}} |\phi_j(r')|^2 \, d^3r' \phi_i(r),$$

$$\hat{K}_i \phi_i(r) = \sum_{j \neq i} \int \frac{e^2}{4\pi\epsilon_0 r_{ij}} \phi_j(r') \phi_i(r') \, d^3r' \phi_j(r).$$

The implementation of HF theory involves solving these Fock equations iteratively in a process known as the self-consistent field (SCF) method. Starting with an initial guess for the molecular orbitals, the Fock matrix is constructed and diagonalized to obtain new orbitals. This procedure is repeated until the molecular orbitals converge, which implies that the input and output orbitals of a cycle are essentially the same, fulfilling the self-consistency condition.

To visualize the convergence of electronic energy and molecular orbitals, the SCF process can be depicted in a convergence plot:

5.6. DENSITY FUNCTIONAL THEORY (DFT)

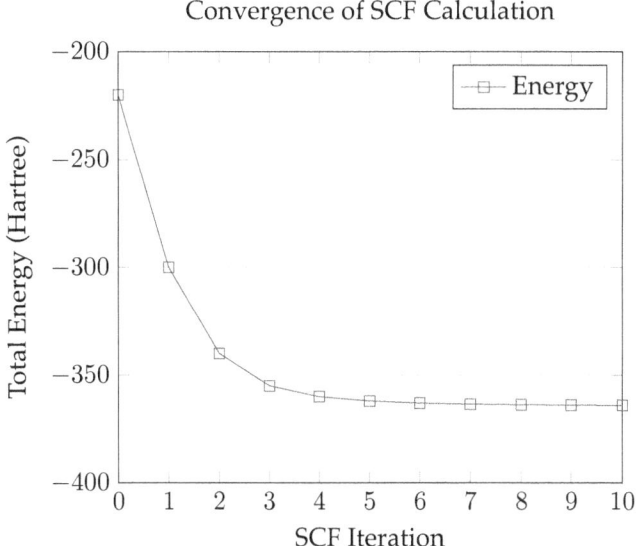

As a result, HF theory provides a framework to not only understand the electronic structure of atoms and molecules but also serves as the foundation for most advanced post-Hartree-Fock methods, which aim to incorporate corrections for electron correlation effects missing in the HF approximation. These subsequent methods assume the solution from HF theory as their starting point, highlighting its pivotal role in the field of quantum chemistry.

5.6 Density Functional Theory (DFT)

Density Functional Theory (DFT) is a quantum mechanical framework enabling the calculation of electronic properties of atoms, molecules, and solids by using functionals—i.e., functions of functions. These functionals map the electron density, a three-dimensional function, to the ground-state energy of the system. DFT circumvents the need to solve Schrödinger's equation for many-electron systems directly, which is computationally prohibitive except for the simplest systems.

The fundamental theorem underpinning DFT is the Hohenberg-Kohn theorem, which states that the ground-state properties of a system are uniquely determined by its electron density. This implies that the total energy of the system can be functionally dependent on the electron density. Consequently, once the electron density that minimizes the total energy is found, all properties of the system can be derived from it. Mathematically, this can be written as:

$$E[\rho] = T[\rho] + V_{\text{ext}}[\rho] + J[\rho] + E_{\text{xc}}[\rho]$$

where $E[\rho]$ is the total energy, $T[\rho]$ is the kinetic energy of the electron system, $V_{\text{ext}}[\rho]$ is the potential energy due to external potentials (usually the nuclei), $J[\rho]$ represents the Coulombic repulsion among the electrons, and $E_{\text{xc}}[\rho]$ is the exchange-correlation energy functional which encapsulates all non-classical interactions between electrons.

The Kohn-Sham system, devised to practically implement the Hohenberg-Kohn concept, postulates that it is possible to replace the many-body problem with a set of non-interacting particles that produce the same density as the interacting system. The Kohn-Sham equations, which are derived from this assumption, resemble a set of Schrödinger equations for a fictitious system of non-interacting electrons:

$$\left(-\frac{\hbar^2}{2m}\nabla^2 + V_{\text{KS}}[\rho]\right)\psi_i = \epsilon_i \psi_i$$

where ψ_i are the Kohn-Sham orbitals, ϵ_i are their corresponding eigenvalues, and $V_{\text{KS}}[\rho]$ is the Kohn-Sham potential. The Kohn-Sham potential itself is given by:

$$V_{\text{KS}}[\rho](\mathbf{r}) = V_{\text{ext}}(\mathbf{r}) + \int \frac{\rho(\mathbf{r}')}{|\mathbf{r} - \mathbf{r}'|} d^3 r' + V_{\text{xc}}[\rho](\mathbf{r})$$

where the first term represents the external potential, the second term is the classical Hartree potential describing electron-electron repulsion, and the third term, $V_{\text{xc}}[\rho]$, is the functional derivative of the exchange-correlation energy with respect to the density.

The practical application of DFT rests on the ability to approximate V_{xc} and E_{xc}. Various approximations have been developed, categorized generally into the local density approximation (LDA), which assumes that the exchange-correlation energy of a point in space depends only on the electron density at that point, and the generalized gradient approximations (GGA), which take into account the gradient of the electron density as well. More sophisticated treatments include meta-GGA and hybrid functionals, which mix in some proportion of exact exchange described by Hartree-Fock theory.

In applying DFT to real systems, the choice of the functional and the treatment of the exchange-correlation effect are critical. Different functionals can lead to significantly different predictions pertaining to electronic properties and ground state geometries, which have to be balanced against computational expense.

Despite these challenges, DFT remains a widely adopted method in computational chemistry, materials science, and physics due to its favorable balance between accuracy and computational cost. As computational resources continue to improve and functionals become more refined, DFT's role in the prediction and understanding of material properties is likely to expand further.

5.7 Semi-empirical and Empirical Methods

The purpose of semi-empirical and empirical methods in quantum chemistry is to simplify the calculations required to predict molecular properties, thus enabling the analysis of larger molecules and complex systems within a reasonable computational time frame. These methods make certain assumptions and use approximations in the electronic Hamiltonian, thus leading to a significant reduction in the computational cost compared to first-principles methods like Hartree-Fock or Density Functional Theory (DFT).

Empirical Methods: Empirical quantum chemical methods utilize experimental data to calibrate their models, often bypassing detailed electronic structure calculations. These methods are generally used for specific types of calculations such as prediction of spectral properties, ionization energies, or molecular geometries. A well-known example of an empirical method is the Hückel Molecular Orbital (HMO) theory, which is specifically designed for the evaluation of π-electron systems in conjugated hydrocarbons. HMO theory simplifies the calculation by considering only π electrons and assumes that the σ electrons contribute a constant background potential. The

CHAPTER 5. QUANTUM MECHANICAL MODELS AND APPROXIMATIONS

Hamiltonian matrix in HMO can be expressed by:

$$\mathbf{H} = \begin{bmatrix} \alpha & \beta & 0 & \cdots & 0 \\ \beta & \alpha & \beta & \cdots & 0 \\ 0 & \beta & \alpha & \cdots & 0 \\ \vdots & \vdots & \vdots & \ddots & \beta \\ 0 & 0 & 0 & \beta & \alpha \end{bmatrix}$$

where α is the Coulomb integral and β is the resonance integral, both adjusted to fit experimental data.

Semi-empirical Methods: Semi-empirical methods are a compromise between empirical approaches and ab initio methods. They incorporate both experimental results and theoretical insight into the electronic interactions. Common semi-empirical models include Austin Model 1 (AM1), Parametric Method 3 (PM3), and the Modified Neglect of Differential Overlap (MNDO). For instance, in the MNDO method, the electronic Hamiltonian is simplified as follows:

$$H_{\text{elec}} = -\sum_i \frac{\nabla_i^2}{2} - \sum_{i,A} \frac{Z_A}{r_{iA}} + \sum_{i<j} \frac{1}{r_{ij}} + \sum_{A<B} \frac{Z_A Z_B}{R_{AB}}$$
$$+ \text{ approximate two-electron integrals}$$

Semi-empirical methods simplify the two-electron integrals using empirical parameters, which are typically fitted to reproduce experimental data or high-quality quantum chemical computations.

To visualize the impact of these simplifications, consider the computational demand involved in calculating the properties of a molecule like acetone (C_3H_6O) using different methods. Figure 5.1 illustrates the computational times required for different levels of chemical accuracy and method types.

Semi-empirical methods can thus handle larger molecular systems and provide reasonably accurate results for a large variety of chemical properties, though they do suffer from lacking theoretical rigor and universal applicability characteristic of more fundamental methods. These methods often become the tools of choice in large-scale computational studies, where a balance between accuracy and computational efficiency must be maintained.

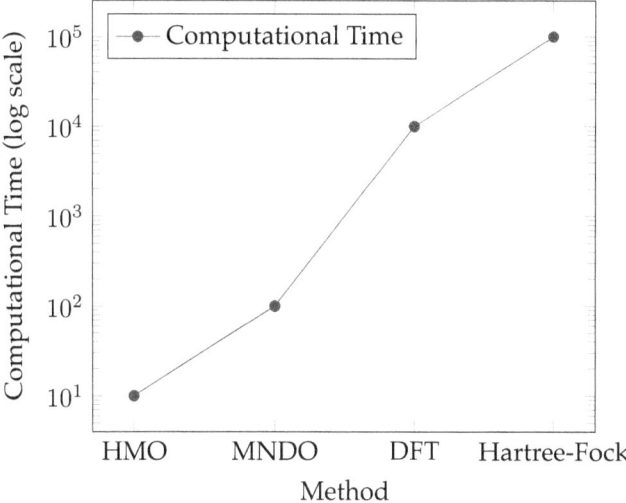

Figure 5.1: Computational time comparison for different quantum chemistry methods for acetone.

5.8 Monte Carlo Simulations in Quantum Chemistry

Monte Carlo simulations represent a stochastic approach to solve physical and mathematical problems by employing random sampling methods to obtain numerical results. In the context of quantum chemistry, Monte Carlo methods provide valuable insights into the behavior of molecular systems, particularly those that are computationally burdensome by deterministic methods like those discussed earlier, such as Hartree-Fock and Density Functional Theory.

The core idea behind Monte Carlo simulations in quantum chemistry is the use of random numbers to sample the positions of particles or the paths of molecules in order to calculate properties such as energy, electron density, or other observables. The advantage of this approach lies in its ability to handle large systems and to approximate the many-body problem with reasonable computational resources.

Quantum Monte Carlo (QMC) methods are a subclass of Monte Carlo methods designed specifically for quantum systems. They are particularly noted for their ability to treat electron correlation more effectively than many other methods. QMC can be subdivided into several types, including Variational Monte Carlo (VMC), Diffusion Monte Carlo (DMC), and Path Integral Monte Carlo (PIMC), each with its specific use case and strength.

Variational Monte Carlo (VMC)

Variational Monte Carlo utilizes a trial wave function, which approximates the ground state wave function of the system. This trial function must be cleverly chosen to ensure that it captures the essential physics of the problem while remaining computationally manageable. The expectation value of the Hamiltonian, relative to this trial wave function, offers an estimate for the ground state energy, derived from:

$$E_{\text{VMC}} = \frac{\int \Psi_T^*(\mathbf{r}) \hat{H} \Psi_T(\mathbf{r}) d\mathbf{r}}{\int |\Psi_T(\mathbf{r})|^2 d\mathbf{r}},$$

where $\Psi_T(\mathbf{r})$ is the trial wave function, $\hat{H}$ is the Hamiltonian, and $\mathbf{r}$ represents the coordinates of the electrons. The integral is calculated using Monte Carlo integration, where configurations of $\mathbf{r}$ are sampled according to $|\Psi_T(\mathbf{r})|^2$.

Diffusion Monte Carlo (DMC)

DMC aims to refine the results from VMC by solving the Schrödinger equation in imaginary time, effectively filtering out higher energy states to converge upon the ground state. The process models the diffusion of probability density, which makes it well-suited for finding the lowest energy state of the system. The DMC energy is given by:

$$E_{\text{DMC}} = \lim_{\tau \to \infty} \frac{\langle \Psi(\mathbf{r}, \tau) | \hat{H} | \Psi_T(\mathbf{r}) \rangle}{\langle \Psi(\mathbf{r}, \tau) | \Psi_T(\mathbf{r}) \rangle},$$

where $\Psi(\mathbf{r}, \tau)$ is the wave function at imaginary time τ.

Path Integral Monte Carlo (PIMC)

PIMC is used to simulate systems at finite temperatures, by taking into account thermal fluctuations. It is based on the path integral formulation of quantum mechanics, where the trajectories of particles are not fixed lines but resemble a cloudy tube of possible paths.

The paths are sampled to evaluate the partition function and other thermal properties, allowing the examination of quantum statistical effects in many-body systems.

Application in Quantum Chemistry: Monte Carlo methods are employed in studying electron correlation in molecules, phase transitions in solids, structure of nanomaterials, and properties of quantum fluids. Especially, they are crucial in cases where other methods fail to provide accurate or feasible solutions due to the sheer size of Hilbert space or complexities involving many-body interactions.

Despite their statistical nature and the associated computational expense, Monte Carlo methods in quantum chemistry have become an essential tool for providing insights into electron behavior and molecular dynamics that are often inaccessible by other means.

5.9 Molecular Dynamics in Quantum Models

Molecular dynamics (MD) simulations provide a dynamic view of molecular systems, which is essential for understanding processes that involve temporal changes, such as chemical reactions, enzyme activity, and other biological mechanisms. In the context of quantum chemistry, quantum molecular dynamics (QMD) simulations play a pivotal role in merging classical dynamical principles with quantum mechanical effects, offering a comprehensive way to study systems that are otherwise computationally prohibitive.

QMD simulations combine classical Newtonian mechanics to describe the overall motion of the atoms in a molecule with quantum mechanics used to treat electron behavior and interactions. The primary challenge in QMD is the computational cost associated with solving the Schrödinger equation for electrons at each timestep in the simulation. There are multiple approaches to implement QMD, each offering different compromises between accuracy and computational feasibility.

Born-Oppenheimer Molecular Dynamics (BOMD) is a traditional approach where the electrons are assumed to be in their ground state for each atomic configuration throughout the trajectory. This method relies heavily on the Born-Oppenheimer approximation, where the electronic and nuclear degrees of freedom are separated. The electronic structure at each step is calculated using methods such as Hartree-Fock or Density Functional Theory. The key equation representing the nuclear motion can be written as:

$$M_I \ddot{\vec{R}}_I = -\nabla_{R_I} E(\vec{R})$$

where M_I is the mass of the Ith nucleus, $\vec{R}_I$ is its position, $\ddot{\vec{R}}_I$ is its acceleration, and $E(\vec{R})$ is the potential energy surface defined by the electronic state.

Car-Parrinello Molecular Dynamics (CPMD) integrates the motion of electrons and nuclei on the fly without necessarily adhering fully to the Born-Oppenheimer approximation. This method introduces a fictitious dynamic for the electronic degrees of freedom, treated as dynamical variables similar to the nuclei. CPMD uses a Lagrangian approach as:

$$\mathcal{L} = \sum_I \frac{1}{2} M_I \dot{\vec{R}}_I^2 + \sum_i \frac{1}{2} \mu \dot{\psi}_i^2 - E(\vec{R}, \{\psi\}) + \sum_{i,j} \Lambda_{ij} (\langle \psi_i | \psi_j \rangle - \delta_{ij})$$

where μ represents the fictitious electron mass, ψ_i are the electronic wavefunctions, and Λ_{ij} are Lagrange multipliers enforcing orthonormality constraints on the wavefunctions.

Unlike classical MD, QMD's inclusion of electronic structure calculations enables the exploration of electronic response to the nuclear movement, making it possible to investigate phenomena such as charge transfer, electronic excitation, and bond breaking/forming during the course of a simulation.

To understand the role of these simulations in practical applications, consider the case of enzyme catalysis. QMD can model the interaction between an enzyme and substrate, detailing how changes in atomic positions influence the electronic structure and, consequently, the reaction mechanism. These insights are pivotal in designing catalysts and understanding disease mechanisms at a molecular level.

Although insightful, QMD is computationally demanding due to the necessity of frequent electronic structure calculations and the management of quantum wavefunctions for a potentially large number of atoms. Recent advancements in computational power, alongside algorithmic developments such as linear scaling methods and the use of graphical processing units (GPUs), are gradually overcoming these limitations, making QMD simulations more accessible for extensive biological and material systems.

Lastly, the fidelity of a QMD simulation to real-world phenomena heavily relies on the quality of the underlying quantum mechanical method and the parameters used in the calculation, such as basis sets and exchange-correlation functionals (in DFT). Thus, it's crucial to carefully choose and test these components based on the specific requirements and sensitivities of the studied system.

Through QMD, quantum chemistry extends its reach to not only static properties of molecules but also their behavior over time, providing a deeper understanding of dynamic processes at the atomic scale.

5.10 Advantages and Limitations of Computational Approximations

As the discourse progresses to encapsulate computational approximations deployed within quantum chemistry, one must delineate both the strengths and constraints associated with these techniques. By scrutinizing these aspects, scientists and researchers are better equipped to select the most appropriate method for particular research questions and scenarios, thereby optimizing both the accuracy and efficiency of their computational experiments.

Advantages of Computational Approximations

1. Enablement of Complex System Analysis In the ambits of quantum chemistry, the sheer complexity and scale of molecular systems often exceed analytical solution capabilities. Computational approximations allow for the study of extensive systems by reducing the problem into manageable elements without the need for exact, often intractable solutions. Methods like Density Functional Theory (DFT) and Hartree-Fock (HF) provide frameworks that balance computational feasibility with reasonable accuracy.

2. Cost-Effectiveness and Resource Allocation Computational methods offer a cost-effective alternative to experimental approaches, especially in the preliminary stages of research where theoretical insights guide experimental design. These techniques reduce the need for expensive laboratory setups and consumables, thus directing resources more judiciously across the research and development spectrum.

3. Facilitation of Predictive Capabilities and Virtual Experiments
Through computational simulations, chemists can predict molecular behavior under a variety of conditions which may not be easily replicated in a laboratory setting. For instance, simulations at extreme temperatures or pressures, toxic or radioactive environments are facilitated via computational models, broadening the understanding of molecular dynamics in unconventional settings.

Limitations of Computational Approximations

1. Reliance on Approximations and Assumptions While indispensable for managing complex calculations, the reliance on approximations can sometimes result in deviations from exact solutions. Accuracy is often sacrificed for feasibility, particularly in methods like the Hartree-Fock, where electron correlation is neglected. Each computational method is underpinned by a set of assumptions, the validity of which governs the accuracy of the results. Recognizing and understanding these limitations is critical when interpreting results.

2. Computational Resource Intensity Despite being cost-effective compared to experimental methods, computational simulations can be resource-intensive, requiring significant high-performance computing power, especially for large systems or highly accurate calculations. The scalar demand for computing power typically results in increased costs and can limit the accessibility of high-level computations, particularly in resource-constrained environments.

3. Accuracy vs. Cost Trade-Offs High-level methods like post-Hartree-Fock techniques offer improved accuracy but at an exponential rise in computational demand. Choosing between accuracy and computational feasibility becomes a critical decision point, particularly in resource-limited projects. Quantum chemists must often negotiate between using a less demanding, less accurate method or a more exact, computationally expensive technique.

4. Lack of Universality No single computational approach is universally valid or applicable. Each method has its realm of applicability, with its accuracy varying depending on the system properties and the specificities of the chemical scenario. For example, DFT performs remarkably well for ground state energies of large systems but may not be as accurate for excited states or systems where dispersion interactions dominate.

While computational approximations offer powerful tools for understanding and predicting chemical phenomena, their effective use necessitates a deep understanding of both their strengths and limitations. Quantum chemists must continuously balance the trade-offs between accuracy, computational demands, and practical feasibility to harness the full potentials of these methodologies. Each computational choice embodies a strategic decision that impacts the quality and reliability of the outcomes, highlighting the pivotal nature of theoretical knowledge in computational chemistry realms.

Chapter 6

Computational Methods in Quantum Chemistry

This chapter outlines the diverse array of computational techniques employed in quantum chemistry to understand and predict molecular properties and behaviors. It elaborates on ab initio methods, which derive properties from basic quantum theory, discusses the significance of semi-empirical methods that use experimental data, and explores density functional theory which has become instrumental in quantum chemical calculations. Additional sections cover quantum Monte Carlo methods and the use of software tools that facilitate these complex calculations, providing insights into how computational quantum chemistry serves as a pivotal tool in modern chemical research and industry applications.

6.1 Introduction to Computational Quantum Chemistry

Computational quantum chemistry is an invaluable branch of chemistry that utilizes quantum mechanical principles to solve chemical problems, allowing for the prediction and explanation of the structure, reactivity, and properties of molecules. This field integrates theories from quantum mechanics with computational methodologies to provide a microscopic view of molecules which is often impossible to obtain through experimental approaches alone.

Quantum chemistry calculations are based on the fundamental assumption that the molecular Schrödinger equation can describe all molecular properties. However, as exact solutions to the Schrödinger equation are only possible for the simplest systems, approximations and computational methods become necessary for more complex molecules.

The earliest approaches to computational quantum chemistry began with the Hartree-Fock method, a mean-field approach where each electron is subject to an average field created by all other electrons. Its equation is formulated as:

$$\hat{F}\psi_i = \epsilon_i \psi_i$$

where $\hat{F}$ is the Fock operator, ψ_i is the wave function of the i-th electron, and ϵ_i represents its energy. This method, however, neglects electron correlation which is crucial for accurate descriptions of molecular properties.

To address this, post-Hartree-Fock methods such as Configuration Interaction (CI) and Møller-Plesset Perturbation Theory (MP2) were developed. These methods involve more sophisticated treatments of electron correlation and can, therefore, predict molecular properties with improved accuracy.

A significant part of computational quantum chemistry is choosing the appropriate level of theory and computational techniques depending on the property of interest, computational resources, and the required accuracy. These decisions can heavily influence the accuracy and reliability of the results.

- **Ab initio methods** - These methods rely entirely on fundamental theories without incorporating experimental data, providing high-accuracy results but often at a significant computational cost.

- **Semi-empirical methods** - These involve some empirical parameters and are generally faster but less accurate than ab initio methods. They are suitable for larger systems where ab initio calculations are computationally prohibitive.

- **Density Functional Theory (DFT)** - This approach has gained popularity due to its good balance between accuracy and computational demands. DFT treats electron density as the primary variable rather than wavefunction.

An overarching consideration in computational quantum chemistry is the trade-off between accuracy and computational expense. Higher accuracy methods demand higher computational resources and time, making them impractical for very large systems or for routine use in high-throughput settings.

Additional complexity arises with the need to correctly manage the basis set, which is a set of functions used to describe the electron orbitals. The choice of basis set can dramatically impact the results of quantum chemical calculations, and is thus a critical component of computational studies.

In recent years, advancements in hardware and software have significantly extended the capabilities of computational chemistry. Parallel computing, in particular, allows for the practical application of computationally expensive methods to larger molecular systems. Software tools that facilitate these calculations are therefore an essential topic of discussion in computational quantum chemistry, bridging the gap between theoretical developments and practical applications.

As this field continues to evolve, computational quantum chemistry remains at the forefront of chemical research, providing insights that are indispensable for the development of new materials, drugs, and understanding chemical phenomena at the molecular level.

6.2 Ab Initio Methods: Basis Sets and Electron Correlation

Ab initio methods, also known as first principles methods, are a cornerstone of quantum chemistry, which employ quantum mechanical principles to compute molecular properties accurately. These methods solve the Schrödinger equation for molecular systems without relying on experimental data, thus providing a predictive framework for understanding molecular behavior under various conditions.

Basis Sets: The choice of basis sets is fundamental in ab initio calculations and directly impacts the accuracy and computational cost. A basis set in quantum chemistry is a set of functions (basis functions) used to approximate the wave function of the electrons in a molecule.

The complexity and type of these functions determine the precision of the results and the resources required for computation.

Commonly used basis sets include:

- STO-nG: Slater-type orbitals approximated by n Gaussians.

- Pople's basis sets: Includes 3-21G, 6-31G, and others, where the numbers signify the complexity and flexibility of the basis functions.

- Correlation-consistent basis sets: cc-pVDZ, cc-pVTZ, etc., designed to systematically converge to the complete basis set limit.

- Plane wave basis sets: Used primarily in periodic systems and materials.

Each type of basis set has its advantages and limitations. For instance, while STO-nG sets are computationally less demanding, correlation-consistent sets, though more expensive, provide higher accuracy, especially in correlation energy calculations.

Electron Correlation: Electron correlation refers to the interaction between electrons in a molecular system which is beyond the mean field approximation of Hartree-Fock theory. Accurately accounting for electron correlation is crucial for reliable predictions of molecular properties such as bond energies, reaction barriers, and excited state energies.

Methods to include electron correlation can be broadly classified into:

- Configuration Interaction (CI): Where the wavefunction is expressed as a linear combination of Slater determinants. CI methods can be exhaustive, considering all possible configurations (Full CI), or truncated to include only single and double substitutions (CISD).

- Many-Body Perturbation Theory (MBPT): Often referred to as Møller-Plesset perturbation theory. MP2, the second-order perturbation, includes interactions between pairs of electrons and is relatively inexpensive.

- Coupled Cluster (CC): This includes the famous CCSD (Coupled Cluster with Single and Double substitutions) and its extension CCSD(T), which includes a perturbative treatment of triple excitations. Coupled cluster methods are highly accurate and often considered the gold standard for quantum chemical calculations.

Both the choice of basis set and the method for including electron correlation are dependent on the trade-off between computational cost and the accuracy required for the study.

Example: Consider a molecule such as methane (CH_4). An ab initio calculation might start with a basic Hartree-Fock calculation using a simple basis set like 3-21G. Subsequently, to improve accuracy, one might employ a CCSD(T) calculation with a more comprehensive basis set like cc-pVTZ to account for electron correlation effectively.

To illustrate the effect of basis set and electron correlation method on the accuracy of molecular properties, consider the calculation of the dissociation energy of a diatomic molecule. Table 6.1 shows hypothetical results using different methods and basis sets.

Method	Basis Set	Dissociation Energy (eV)
HF	3-21G	3.2
HF	cc-pVTZ	3.35
CCSD	3-21G	4.1
CCSD(T)	cc-pVTZ	4.5

Table 6.1: Hypothetical dissociation energies for a diatomic molecule calculated using different methods and basis sets.

The table suggests the importance of both an advanced correlation method and a higher-level basis set to achieve more accurate results. Furthermore, understanding these concepts allows researchers and students to digitally simulate potential novel compounds and predict properties, streamlining the experimental process and guiding scientific exploration in molecular design and discovery.

6.3 Semi-Empirical Methods: Parameters and Applications

Semi-empirical methods stand as a critical facet of computational quantum chemistry, providing a practical balance between computational efficiency and accuracy. These methods are derived from quantum mechanical principles but involve empirically derived parameters to simplify calculations. The key principle behind semi-empirical methods is that they use a combination of theoretical quantum mechanics and experimental data to approximate solutions to the Schrödinger equation.

The Foundation of Semi-Empirical Methods

The foundation of semi-empirical methods lies in the Molecular Orbital (MO) theory. In MO theory, the electronic structure of molecules is described by linear combinations of atomic orbitals (LCAO). The electronic Hamiltonian, which encapsulates the total energy of electrons in a molecule, is thus approximated and solved for these molecular orbitals. Semi-empirical methods simplify the computation-intensive steps of ab initio methods by using empirical data to preset certain integrals, particularly those involving electron correlation and electron repulsion, which are otherwise computational costly.

Hamiltonian and Parameterization

In semi-empirical calculations, the Hamiltonian retains terms crucial for defining electron behavior but neglects or approximates others through parameters derived from experimental results. For instance, terms describing core-core electron repulsions can be parameterized as they change minimally between similar types of molecules.

Mathematically, the Hamiltonian in semi-empirical methods can be represented as:

$$\hat{H} = -\sum_i \frac{\hbar^2}{2m}\nabla_i^2 - \sum_{i,j} \frac{Z_j e^2}{4\pi\epsilon_0 r_{ij}} + \sum_{i<j} \frac{e^2}{4\pi\epsilon_0 r_{ij}} + \text{other terms}$$

where $\hbar$ is the reduced Planck's constant, m is the electron mass, Z_j are the atomic numbers, e is the elementary charge, and ϵ_0 is the permittivity of free space. The electron-electron repulsion and electron-nuclear attractions are explicitly considered, yet many-body terms and certain integrals involving these are parameterized.

6.3. SEMI-EMPIRICAL METHODS: PARAMETERS AND APPLICATIONS

Common Semi-Empirical Models

Several semi-empirical models such as the Huckel method, Extended Huckel Theory (EHT), CNDO (Complete Neglect of Differential Overlap), INDO (Intermediate Neglect of Differential Overlap), and NDDO (Neglect of Diatomic Differential Overlap) have been developed. Each of these models differ in how the approximations are applied and the specific parameters used.

- **Huckel method**: Mainly used for conjugated hydrocarbon systems, it considers only π electrons and assumes zero overlap between non-neighboring atomic orbitals.

- **Extended Huckel Theory (EHT)**: An advancement over the Huckel method; it includes both σ and π electrons and allows for overlap and considers more integrals but still employs significant approximations in electron interactions.

- **CNDO/INDO/NDDO**: These methods offer increased sophistication allowing a better treatment of electron repulsion by using different levels of neglect for different overlap integrals. NDDO, for example, which is used in popular semi-empirical packages like AM1 and PM3, include parameters fitted to a large set of experimental data.

Applications of Semi-Empirical Methods

Applications of semi-empirical methods are vast due to their reduced computational demands compared to fully ab initio calculations. These methods find utility in:

- Modelling large molecules and molecular complexes which would be computationally prohibitive with ab initio methods.

- Preliminary screenings in drug discovery processes to identify potential drug candidates by predicting activities, binding affinities, or reactivity.

- Material science applications for modeling polymers and organic materials where detailed electron correlation is less critical.

Moreover, semi-empirical methods are indispensable in educational settings and initial phase research where quick and reasonably accurate estimates of molecular properties are required without the extensive resource allocation necessary for more rigorous quantum mechanical methods.

In summary, while semi-empirical methods make considerable approximations, their calibrated approach leveraging both theoretical constructs and empirical data provides a valuable tool in the arsenal of computational chemistry, balancing the scales of accuracy and computational feasibility.

6.4 Density Functional Theory (DFT): Foundations and Functional

Density Functional Theory (DFT) constitutes a major paradigm within the quantum chemistry landscape, enabling the treatment of several-body systems through the use of functionals—mathematical functions of another function. The cornerstone of DFT lies in the Hohenberg-Kohn theorems, which established the theoretical foundation of this approach by proving that the ground state properties of a many-electron system are uniquely determined by the electron density, $\rho(\vec{r})$.

The primary equation in DFT is the Kohn-Sham equation, which effectively transforms the many-body problem into a series of single-particle problems. This equation is expressed as:

$$\left[-\frac{1}{2}\nabla^2 + V_{\text{eff}}(\vec{r})\right] \phi_i(\vec{r}) = \epsilon_i \phi_i(\vec{r}),$$

where $\phi_i(\vec{r})$ are the Kohn-Sham orbitals, and ϵ_i their corresponding eigenvalues. The effective potential, $V_{\text{eff}}(\vec{r})$, is a sum of the external potential, $V_{\text{ext}}(\vec{r})$, the electron-electron interaction potential, and the exchange-correlation potential, denoted as $V_{\text{xc}}(\vec{r})$. The computation of $V_{\text{xc}}(\vec{r})$ is central to DFT and is derived from the exchange-correlation functional, $E_{\text{xc}}[\rho]$.

The challenge in DFT is largely associated with the accurate approximation of the exchange-correlation functional. The exact form of $E_{\text{xc}}[\rho]$ is unknown and must be approximated. Common approximations take the form of either local-density approximations (LDA) or generalized gradient approximations (GGA). LDA assumes that E_{xc} at each point in space depends only on the density at that point, and is expressed as:

$$E_{xc}^{LDA}[\rho] = \int \epsilon_{xc}(\rho(\vec{r}))\rho(\vec{r})\,d^3r,$$

where $\epsilon_{xc}(\rho)$ is a function derived from the uniform electron gas model.

On the other hand, GGA includes the dependence on the gradient of the density as well, which can be formalized as:

$$E_{xc}^{GGA}[\rho] = \int \epsilon_{xc}(\rho(\vec{r}), \nabla\rho(\vec{r}))\rho(\vec{r})\,d^3r,$$

where $\epsilon_{xc}(\rho, \nabla\rho)$ accounts for inhomogeneities in the electron density.

Dependence of E_{xc} on ρ and $\nabla\rho$

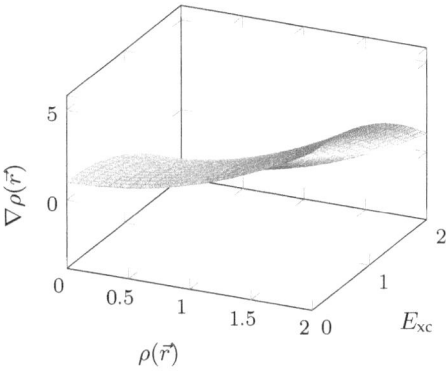

This differential treatment between LDA and GGA forms the basis of much of the practical implementations of DFT. Advanced functionals, such as hybrid functionals, further improve accuracy by mixing GGA with elements of Hartree-Fock theory, such as the exact exchange energy. These developments continue to expand DFT's utility in tackling increasingly complex systems, proving indispensable in materials science, molecular biology, and nanotechnology.

6.5 Quantum Monte Carlo Methods

Quantum Monte Carlo (QMC) methods constitute a class of stochastic approaches designed to solve the quantum many-body problem. Unlike deterministic methods such as Hartree-Fock or Density Functional Theory, QMC tackles the electron correlation problem by employing statistical sampling of the wave function squared, $|\Psi|^2$, which corresponds to the probability density of finding electrons in certain configurations.

Basics of Quantum Monte Carlo: At the core of QMC is the representation of the quantum state of a system using a set of random walkers, which are sampled according to a distribution proportional to the quantum-mechanical probability density. The inherent randomness in this approach is akin to typical Monte Carlo methods used in other fields of physics and computational science.

The fundamental equation governing QMC methods is the Schrödinger equation:

$$\hat{H}\Psi = E\Psi$$

where $\hat{H}$ is the Hamiltonian operator, Ψ is the wave function of the system, and E is the energy eigenvalue associated with Ψ. The goal is to approximate Ψ in such a way that the expectation value of $\hat{H}$, and thus E, can be accurately computed.

Variants of QMC: There are several types of QMC methods, each with its unique strategies and applications:

1. *Variational Monte Carlo (VMC)* – In VMC, a trial wave function $\Psi_T(\vec{R};\theta)$, parameterized by variables θ, is used. The parameters are adjusted to minimize the expectation value of the energy:

$$E_{VMC} = \frac{\int \Psi_T^*(\vec{R};\theta)\hat{H}\Psi_T(\vec{R};\theta)d\vec{R}}{\int |\Psi_T(\vec{R};\theta)|^2 d\vec{R}}$$

The minimization process typically involves optimization algorithms such as steepest descent or genetic algorithms.

2. *Diffusion Monte Carlo (DMC)* – This method allows the projection of a trial wave function onto the ground state wave function by simulating a diffusion process mixed with a branching process of replication or deletion of sample points (walkers). DMC is highly effective for ground state studies but requires a good trial wave function, generally provided by VMC, to initiate the simulation.

3. *Path Integral Monte Carlo (PIMC)* – PIMC is used to study finite-temperature properties of quantum systems. It involves sampling over paths or trajectories in the imaginary time, representing the partition function of the quantum system.

Implementation and Challenges: The execution of QMC methods is computationally demanding, requiring efficient algorithms and sometimes paralleled or distributed computing environments. The methods face several challenges, including the fermion sign problem associated with antisymmetric fermionic wave functions leading to potential cancellations in the Monte Carlo sampling process, known as 'sign problem'.

Applications: Despite these challenges, QMC methods have been successfully applied in calculating properties of atoms, molecules, and condensed matter systems. For example, the ground-state energies, excited state properties, and electron correlation effects are accessible through accurate QMC simulations.

Comparison with Other Methods: QMC is often considered superior to other quantum computational methods in dealing with the electron-electron correlation problem directly. However, the computational cost and the difficulties associated with the fermion sign problem often limit its application to relatively smaller systems or when high accuracy is essential.

Future Prospects: Ongoing developments in algorithmic strategies and computational power are expected to expand the applicability and accuracy of QMC methods. Adaptations like the use of machine learning techniques for optimizing trial wave functions or hybrid approaches integrating DFT or semi-empirical methods with QMC are current areas of research, promising to enhance capabilities and efficiency.

The choice of QMC method and its specific implementation can significantly influence the accuracy and computational feasibility of quantum chemical calculations, illustrating the importance of these methods in computational chemistry toolbox.

6.6 Molecular Mechanics Versus Quantum Mechanics

In the domain of computational chemistry, the methodologies adopted for investigating molecular systems generally fall into two categories: molecular mechanics and quantum mechanics. Each approach carries its own set of assumptions, methodologies, and applications, making them suitable for different kinds of problems.

This section aims to delineate these differences, drawing clear contrasts and highlighting the scenarios in which one might be preferred over the other.

Foundations of Molecular Mechanics

Molecular mechanics (MM) applies classical physics to model molecular systems. The primary principle behind MM is that the total energy of a molecule can be described as a function of the positions of all atoms in the molecule. This is often expressed as a potential energy function or force field, comprising terms that describe bond stretching, angle bending, torsional angles, non-bonded interactions such as van der Waals forces, and electrostatic charges.

The mathematical representation of the potential energy, V, in molecular mechanics is:

$$V = V_{bond} + V_{angle} + V_{torsion} + V_{non-bonded}$$

where:

- V_{bond} represents the energy due to bond stretching
- V_{angle} is the energy contribution from bond angle distortions
- $V_{torsion}$ involves torsional or dihedral angle distortions
- $V_{non-bonded}$ encompasses van der Waals and electrostatic interactions

MM is particularly effective in studying large biological molecules and material assemblies where quantum mechanics calculations become computationally prohibitive. The simplifications in molecular mechanics, however, come at the cost of not explicitly considering electronic effects which are often crucial in chemical reactivity and properties.

Principles of Quantum Mechanics

Quantum mechanics (QM), unlike molecular mechanics, inherently considers the electronic structure of molecules. This field operates on the fundamental premise that the properties of atoms and molecules are governed by the Schrödinger equation—a differential equation

6.6. MOLECULAR MECHANICS VERSUS QUANTUM MECHANICS

that describes how the quantum state of a physical system changes with time.

For a molecular system, the stationary form without considering time-dependence is given by:

$$\hat{H}\Psi = E\Psi$$

where $\hat{H}$ is the Hamiltonian operator symbolizing the total energy of the system, Ψ represents the wave function of the system, and E is the energy eigenvalue associated with Ψ. Quantum mechanics provides a detailed description of the electronic structure through solutions to the Schrödinger equation, offering insights into bonding, molecular orbitals, and properties like ionization energy and electron affinity.

Quantum chemical calculations often require substantial computational resources, particularly for large systems and when higher levels of theory are used. However, they are indispensable for accurately predicting and understanding fundamental chemical phenomena and properties.

Comparative Analysis and Application Domain

While molecular mechanics might be adequate for understanding structural properties and dynamics of large biomolecules or materials where fine electronic details are less critical, quantum mechanics is critical when detailed knowledge about electronic distribution is essential, such as in the study of chemical reactivity, photochemistry, and the properties of transition metal complexes.

Furthermore, the choice between quantum mechanics and molecular mechanics can also be influenced by the computational resources available. For extensive systems such as proteins, nucleic acids, or materials, molecular mechanics provides a viable solution. On the other hand, for small to moderately sized molecules, or when dealing with processes where electron movement is crucial (like in excited states or in redox reactions), quantum mechanics is more suitable.

Molecular mechanics and quantum mechanics serve as two fundamental yet distinct methodologies in computational chemistry. The choice between these approaches should be guided by the nature of the problem at hand, the level of electronic detail required, and the computational resources available. Understanding the strengths and limitations of each method facilitates their effective application in chemical research, providing a robust framework for studying a wide array of molecular systems.

6.7 Software and Tools in Quantum Chemistry

In quantum chemistry, the complexity and scale of computations require robust software and specialized tools. These tools are not just facilitators but are essential drivers of research and application in the field. Given the computational demands, these software packages are designed to utilize advanced algorithms for solving the Schrödinger equation for many-electron systems, among other tasks. This section will explore some of the most pivotal software used in quantum chemistry, including Gaussian, ORCA, and NWChem, and how these tools are applied in different scenarios.

Gaussian is perhaps one of the most broadly used software in the field of quantum chemistry. It provides a suite of algorithms covering a wide range of computational chemistry methods including ab initio methods, density functional theory, and semi-empirical techniques. Gaussian is particularly noted for its extensive options for basis sets and integration of functional analytics for DFT calculations. The ability to predict vibrational modes, molecular orbitals, and reaction pathways makes Gaussian a versatile tool for chemists.

```
Example of Gaussian input for a water molecule optimization:
#P B3LYP/6-31G(d) opt

Title Card Required

0 1
H
O 1 r2
H 2 r3 1 a3

r2 = 0.958
r3 = 0.958
a3 = 104.5
```

ORCA is another widely utilized software, known for its high-performance algorithms for electronic structure computations. It supports a variety of methods including DFT, Hartree-Fock (HF), and post-HF methods such as Coupled Cluster (CC) with singles, doubles and perturbative triples [CCSD(T)], which are critical for understanding electron correlation with high accuracy. ORCA also

6.7. SOFTWARE AND TOOLS IN QUANTUM CHEMISTRY

provides tools for calculating spectroscopic properties of molecules, which is crucial for comparing theoretical predictions with experimental results.

NWChem is designed to handle both small to large scale molecular structures and provides capabilities for biomolecular systems as well. It supports methodologies ranging from quantum mechanical calculations to classical molecular dynamics simulations. NWChem's ability to run on high-performance parallel computing environments makes it particularly useful for studying complex systems at a reasonable computational cost.

The use of these software tools is underpinned by the need for efficient computational infrastructure. The calculations involved in quantum chemistry are resource-intensive, involving large matrices and multi-dimensional integrals. For instance, the Hartree-Fock method involves solving the Roothan equations, which are a set of coupled secular equations:

$$FC = SCE$$

where F is the Fock matrix, C the matrix of expansion coefficients, S the overlap matrix, and E the diagonal matrix of orbital energies. The computational cost becomes a significant factor when dealing with basis sets that expand the electron density in multiple Gaussian-type orbitals.

Parallel computing environments used in conjunction with these software tools greatly enhance the capabilities of chemists to simulate and predict molecular behaviors. MPI (Message Passing Interface) and GPU (Graphics Processing Units) acceleration are commonly leveraged technologies. For instance, Gaussian and NWChem both support GPU acceleration for speed-ups in basis set evaluations and electron repulsion integral calculations.

In practical applications, computational chemists often need to interface different software tools depending on the specific requirements of the project. Tools like Molden or ChemCraft are used for visualizing molecular structures and electronic densities, which are often output from calculations performed in Gaussian or ORCA.

To summarize, the software and tools in quantum chemistry are as critical to the field as theoretical foundations. They enable detailed exploration of molecular systems, handle the strenuous computational load, and provide insights that are otherwise infeasible to ob-

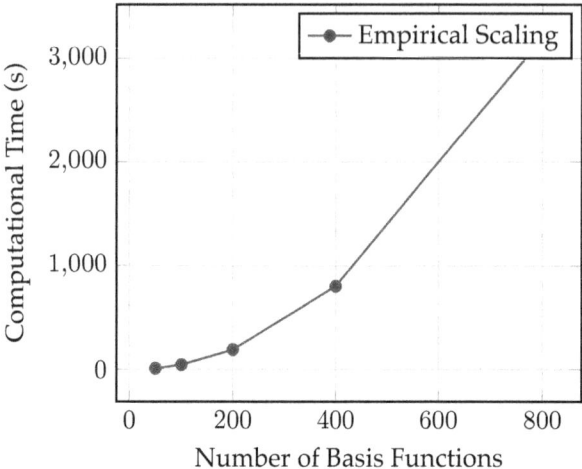

Figure 6.1: An illustrative example showing the scaling of computational time with the increase in the number of basis functions in quantum chemical calculations.

tain, particularly for molecules of industrial and biological significance.

6.8 Quantum Chemistry Simulations: Setup and Execution

Quantum chemistry simulations are a cornerstone for analyzing and predicting molecular behavior using quantum mechanical principles. These simulations are executed through specialized software tools and often require substantial computational resources. The execution of quantum chemistry simulations involves several key steps: system setup, selection of computational methods and parameters, execution of computations, and analysis of results. This section delves into the intricacies of each step, highlighting their importance and impact on the final simulation results.

6.8. QUANTUM CHEMISTRY SIMULATIONS: SETUP AND EXECUTION

System Setup

In quantum chemistry simulations, the initial system setup is critical. It involves defining the molecular system's electronic and geometric structure, which can be either derived from experimental data or computed using molecular modeling tools. For example, a common format for specifying molecular geometry is the XYZ format, in which the atomic species and their coordinates are defined as follows:

```
C    0.0000    0.0000    0.0000
H    0.0000    0.0000    1.0870
H    1.0267    0.0000   -0.3630
H   -0.5135   -0.8892   -0.3630
H   -0.5135    0.8892   -0.3630
```

Once the structure is defined, electronic configurations must also be set up, involving decisions on the type of electrons (e.g., core vs. valence), and any necessary constraints or symmetries in the system.

Selection of Computational Methods and Parameters

The choice of computational methods in quantum chemistry simulations is influenced by the nature of the molecular system and the properties of interest. Key methodologies include Hartree-Fock (HF) theory, post-Hartree-Fock methods, Density Functional Theory (DFT), and others. Each method has its own set of parameters, such as basis sets in HF theory and functionals in DFT, that need to be meticulously chosen to balance accuracy and computational efficiency.

A typical choice of a basis set might be demonstrated

```
\ce{6-31G**}
```

Execution of Computations

The execution of quantum chemistry simulations often requires high-performance computing environments due to the computational demands of quantum mechanical calculations. This step involves submitting job files to computational clusters, which often include directives for resource allocation, executable paths, and method-specific commands. An example of a job file setup using PBS (Portable Batch System) might look like:

```
#PBS -N quantum_simulation
#PBS -l nodes=2:ppn=16
#PBS -l walltime=48:00:00
#PBS -q standard
#PBS -o output_file.txt
#PBS -e error_file.txt

cd $PBS_O_WORKDIR
module load gaussian
g09 < input_file.gjf > output_file.log
```

Analysis of Results

After the simulation, the output files need to be analyzed to extract meaningful chemical insights. This involves reading computational logs to evaluate convergence criteria, extracting energies, electron densities, and other properties, and visualizing molecular orbitals and other quantum mechanical properties using specialized software tools like GaussView or Avogadro.

```
Energy: -40.507 Hartrees
Dipole moment: 1.24 Debye
```

Setting up and executing quantum chemistry simulations is a meticulous process that requires careful planning and attention to detail at every stage. From the initial system setup to the final analysis of results, each step must be executed with precision to ensure the reliability and relevance of the computational predictions.

6.9 Analyzing Quantum Chemistry Computational Results

Understanding the computational results of quantum chemistry simulations is critical for interpreting the physical and chemical properties of molecules. Results from such computations typically include data on electron density, molecular orbitals, energy states, and derived properties like dipole moments and reaction pathways. This section examines how these outcomes are extracted, analyzed, and interpreted using specific examples and notable equations.

To commence, it is pertinent to analyze the outcomes in terms of molecular geometry optimization. Molecular geometries obtained from quantum calculations are typically optimized using gradient descent methods, which iteratively adjust atomic positions to minimize

6.9. ANALYZING QUANTUM CHEMISTRY COMPUTATIONAL RESULTS

the molecular energy. The final geometry is crucial as it provides a foundation for further property calculation. The mathematical representation of the optimization process is often given by:

$$E_{\text{opt}} = \min_{\mathbf{R}} E(\mathbf{R})$$

where $E(\mathbf{R})$ is the energy of the molecule as a function of nuclear coordinates $\mathbf{R}$. The optimization seeks the set of coordinates $\mathbf{R}$ that results in the lowest possible energy E_{opt}.

Following geometry optimization, electronic properties including the electron density and molecular orbitals are computed. One might employ the electron density function, $\rho(\mathbf{r})$, defined as:

$$\rho(\mathbf{r}) = \sum_i |\psi_i(\mathbf{r})|^2$$

where $\psi_i(\mathbf{r})$ are the molecular orbitals. The visualization of electron density can provide insights into electron distribution within the molecule and is critical for analyzing chemical reactivity and bonding properties.

Molecular orbitals are often analyzed using plots derived from computational simulations. These plots are developed using data that represents the probability density function of electrons in these orbitals. Energy of these orbitals, particularly the Highest Occupied Molecular Orbital (HOMO) and Lowest Unoccupied Molecular Orbital (LUMO), is crucial as it impacts chemical reactivity. The HOMO-LUMO gap, a key descriptor in several chemical applications, is computed as:

$$\Delta E = E_{\text{LUMO}} - E_{\text{HOMO}}$$

Another integral component of computational quantum chemistry results is thermodynamic properties calculation, which includes enthalpy, entropy, and free energy changes. These are often computed based on the statistical mechanics principals integrating over molecular energies from the simulation. For instance, the Gibbs free energy, often calculated in quantum chemistry simulations, is given by:

$$G = H - TS$$

where H represents enthalpy, T the temperature, and S the entropy.

Furthermore, dipole moments, an essential outcome of quantum chemistry calculations, offer insight into the polarity of a molecule and are calculated using the formula:

$$\mu = \int \rho(\mathbf{r})(\mathbf{r} - \mathbf{r}_{\text{com}})\,d^3r$$

where $\mathbf{r}_{\text{com}}$ is the center of mass of the charge distribution.

Finally, reaction pathways and transition states can be analyzed by constructing potential energy surfaces that map energy changes as a function of reaction coordinates. This is crucial for understanding reaction mechanisms and predicting product formation.

Graphically, many of these results are represented using plots and figures. For instance, a common analysis tool includes generating 'energy vs. reaction coordinate' plots, which visually depict potential energy surfaces. In practice, this is done using tools like Gnuplot or Matplotlib in Python, which can plot energies extracted from output files of the quantum chemistry computations.

Analyzing results from quantum chemistry computations involves a detailed investigation of molecular properties and behaviors at an electronic level. It encompasses geometrical optimizations, electronic structure calculations, thermodynamic and kinetic property calculations, and the construction of visual representations for easier understanding and interpretations.

Chapter 7

Spectroscopy and Quantum Chemistry

This chapter delves into the integration of spectroscopy with quantum chemistry, explaining how spectroscopic techniques are indispensable tools for examining and confirming theoretical predictions about molecular and electronic structures. It covers a range of spectroscopic methods including rotational, vibrational, electronic, and magnetic resonance spectroscopies. Each method is discussed in the context of quantum mechanical principles, highlighting how changes at the quantum level affect spectroscopic outcomes. This comprehensive examination provides a clearer understanding of the molecular world through the lens of both experimental and theoretical quantum chemistry.

7.1 Introduction to Spectroscopy in Quantum Chemistry

Spectroscopy, fundamentally, is the study of the interaction between light and matter. This branch of science has evolved into a critical tool for quantum chemists, providing a spectrum-based insight into the structure and behavior of atoms and molecules. Spectroscopic techniques harness the principles of light-matter interaction to probe the electronic states, molecular energies, and dynamic processes, which are the purview of quantum mechanics.

Quantum mechanics underpins the theoretical framework of spectroscopy in explaining how atoms and molecules absorb, emit, and scatter electromagnetic radiation. The quantum mechanical model of the atom, particularly the quantization of energy states, is critical in understanding why spectroscopy works. Each quantum state corresponding to a specific energy level is described by a wavefunction, which adheres to the Schrödinger equation:

$$i\hbar \frac{\partial}{\partial t} \Psi(r,t) = \hat{H} \Psi(r,t)$$

where i is the imaginary unit, $\hbar$ is the reduced Planck's constant, $\Psi(r,t)$ is the wavefunction of the system, and $\hat{H}$ the Hamiltonian operator. The solutions to this equation provide the energy eigenvalues and eigenfunctions, which are crucial in predicting the spectral lines.

In spectroscopy, when photons are absorbed or emitted, transitions occur between these quantized energy levels. The energy of the photon corresponds to the difference between the initial and final energy states ($E = h\nu$), where h is Planck's constant and ν is the frequency of the photon. These transitions give rise to the spectral lines that are characteristic of each element or molecule:

$$\Delta E = E_f - E_i = h\nu$$

Exploring various branches, spectroscopy can be divided fundamentally into absorption, emission, and scattering spectroscopy: - **Absorption Spectroscopy** involves the measurement of the wavelength and intensity of absorption of light by a sample. The absorbed light typically promotes the molecule from a lower to a higher energy state. - **Emission Spectroscopy** involves measuring the light emitted from a sample after it has been excited by heat or light. This is often used to study electronic states in fluorescent and phosphorescent materials. - **Scattering Spectroscopy** examines the light that a sample scatters at various angles.

Each technique explores different attributes of electron configurations and molecular vibrations, pivoting on quantum mechanical interactions:

- **Rotational Spectroscopy** examines the microwave region of the electromagnetic spectrum and probes changes in the rotational motion of molecules, encapsulated effectively by the rotational Hamiltonian:

$$\hat{H}_{rot} = \frac{\hat{J}^2}{2I}$$

where $\hat{J}$ is the angular momentum operator and I is the moment of inertia.

- **Vibrational Spectroscopy (Infrared and Raman)** investigates molecular vibrations and uses infrared light to stimulate transitions between vibrational energy levels, characterized by the vibrational Hamiltonian:

$$\hat{H}_{vib} = \sum_i \frac{1}{2} k x_i^2 + \frac{p_i^2}{2m_i}$$

where k is the force constant, x_i is the displacement from equilibrium, and m_i is the mass of the i-th atom.

- **Electronic Spectroscopy (UV-visible)** involves transitions between electronic energy levels induced by ultraviolet or visible light, crucial for understanding electronic structure and band gaps in materials.

- **Magnetic Resonance Spectroscopies**, including Nuclear Magnetic Resonance (NMR) and Electron Spin Resonance (ESR), investigate the properties of atomic nuclei and electrons in a magnetic field, providing exquisite details about the environment, structure, and dynamics of molecules.

7.2 Principles of Spectroscopic Techniques

The analysis of light as it interacts with matter constitutes the core of spectroscopic techniques, providing invaluable insights into the molecular and electronic architecture of substances. To comprehend how these techniques interlace with quantum chemistry, it is crucial to understand the fundamental principles guiding these interactions. At its core, quantum chemistry provides a theoretical framework that explains the intrinsic nature of matter at the atomic and molecular levels, which is essential for interpreting spectroscopic data.

Spectroscopy capitalizes on the fact that when molecules absorb or emit electromagnetic radiation, they transition between different quantum states. According to quantum mechanics, these states are quantized, with specific energies, momenta, angular momenta, and other physical properties defined by the Schrödinger equation. The spectroscopic techniques, therefore, revolve around measuring the energy differences between these states, which correspond to the energy of the radiation absorbed or emitted during the transition.

Interaction of Radiation with Matter

The first principle to consider is the interaction of radiation with matter. When electromagnetic radiation interacts with matter, it can be absorbed, emitted, or scattered, and each of these interactions can be described using quantum mechanical principles. The fundamental equation governing these processes is the time-dependent Schrödinger equation, given by:

$$i\hbar \frac{\partial}{\partial t} \Psi(r, t) = \hat{H} \Psi(r, t)$$

where i is the imaginary unit, $\hbar$ is the reduced Planck constant, $\Psi(r,t)$ is the wave function of the system, and $\hat{H}$ is the Hamiltonian operator representing the total energy of the system.

At the heart of the interaction between light and matter is the electric dipole moment, which is induced or inherent in a molecule. The selection rules for spectroscopic transitions, which dictate the allowed transitions between energy levels, depend significantly on the symmetry properties of the molecules and the characteristics of the electromagnetic radiation involved.

Energy Transitions

The absorption of electromagnetic radiation leads to an upward transition in energy levels, while emission results in a downward transition. These processes can be succinctly represented as:

$$\Delta E = h\nu$$

where ΔE is the change in energy, h is Planck's constant, and ν is the frequency of the absorbed or emitted radiation.

Transitions between different types of energy levels—electronic, vibrational, and rotational—are the basis for different spectroscopic methods. Each type of spectroscopy focuses on specific kinds of transitions:

- **Rotational Spectroscopy**: Studies transitions between different rotational states of molecules, typically in the microwave region of the electromagnetic spectrum.

- **Vibrational Spectroscopy**: Involves transitions between vibrational states, mostly observed in the infrared or Raman spectra.

- **Electronic Spectroscopy**: Focuses on transitions between electronic states, usually visible in the UV-visible spectra.

- **Magnetic Resonance Spectroscopies**: E.g., NMR and ESR, which detect changes in the spin states of nuclei or electrons.

Techniques and Instrumentation

Each spectroscopic method employs specific types of instrumentation designed to detect and measure the particular type of radiation involved in the transitions of interest. For example, a spectrophotometer is generally used in UV-visible spectroscopy to measure how much light is absorbed by a sample and at what wavelengths. In contrast, a Fourier-transform infrared (FTIR) spectrometer is employed in infrared spectroscopy to record the intensity of light as a function of wavelength, which is mathematically transformed to yield an infrared spectrum.

Quantitative Analysis

The measured spectroscopic data can often be quantitatively analyzed to determine the concentration of substances, the bond strength, or the dynamic processes at play within a sample. Techniques such as Beer-Lambert law in UV-visible spectroscopy provide a direct method to relate the intensity of absorbed light to the concentration of the absorbing species in a sample:

$$A = \epsilon l c$$

where A is the absorbance, ϵ is the molar absorptivity, l is the path length of the light through the sample, and c is the concentration of the absorbing substance.

The integration of these principles in quantum chemistry through the use of spectroscopic techniques allows for a more profound understanding of molecular and electronic structures that are otherwise impossible to observe directly.

7.3 Rotational Spectroscopy: Quantum Mechanical View

Rotational spectroscopy, or microwave spectroscopy, involves the measurement of the radiation absorbed or emitted by molecules as they undergo transitions between rotational energy levels. This section explores the quantum mechanical foundations of rotational spectroscopy, providing insights into how the spectroscopic data relate to molecular structure and dynamics.

Quantum Mechanical Model of Rotational Spectroscopy

Rotational energy levels in molecules arise from their rotational motion, quantized according to quantum mechanics. For a rigid rotor, which is an idealized model of a rotating molecule, the rotational energy levels are given by:

$$E_J = \frac{\hbar^2}{2I} J(J+1)$$

where J is the rotational quantum number ($J = 0, 1, 2, \ldots$), $\hbar$ is the reduced Planck's constant, and I is the moment of inertia of the molecule about the axis of rotation. The moment of inertia I is a crucial quantity and depends on the molecular mass and geometry, defined as:

$$I = \mu r^2$$

where μ is the reduced mass of the molecule and r is the bond length, assuming a diatomic molecule.

Selection Rules and Spectral Transitions

The quantum mechanical selection rules dictate the allowed transitions between rotational energy levels. For rotational spectroscopy, the selection rule is:

$$\Delta J = \pm 1$$

7.3. ROTATIONAL SPECTROSCOPY: QUANTUM MECHANICAL VIEW

This rule means that during the absorption or emission of a photon, the rotational quantum number J can only change by 1. The frequency of the photon corresponding to this transition is given by:

$$\nu = \frac{E_{J+1} - E_J}{h} = \frac{2B(J+1)}{h}$$

where $B = \frac{\hbar^2}{2I}$ is the rotational constant and h is Planck's constant.

Spectral Lines and Molecular Structure

The spectral lines observed in rotational spectroscopy are directly related to the inertia and shape of the molecule. Molecules with higher moments of inertia (larger or heavier molecules) exhibit spectral lines at lower frequencies. From Equation 7.3, the spacing between the spectral lines is constant ($2B$), which simplifies the interpretation of the spectrum. By analyzing these spacings, insights into the molecular structure, including bond lengths and molecular geometry, can be derived.

Effects of Isotopic Substitution

Isotopic substitution is a useful method in rotational spectroscopy to glean more information about molecular structure. Replacing an atom with its heavier isotope affects the moment of inertia, and thus the rotational constant B, causing a shift in the spectral lines. The extent and nature of the shift can be analyzed to provide precise information about the bond where the isotope has been substituted.

Application of Rotational Spectroscopy

Applications of rotational spectroscopy extend into numerous fields including atmospheric chemistry, astrophysics, and molecular identification. For instance, the identification of interstellar molecules is largely based on the analysis of their microwave spectral lines. The precise quantification and complete rotational spectrum enable astronomers to identify molecular species even in distant cosmic environments.

The quantum mechanical view of rotational spectroscopy offers a profound understanding of molecular dynamics and structure.

Through rigorous analysis of the rotational spectra, one can extract detailed structural parameters and dynamics, profoundly impacting various scientific domains.

7.4 Vibrational Spectroscopy: IR and Raman

Vibrational spectroscopy encompasses two primary techniques of profound relevance in quantum chemistry: infrared (IR) and Raman spectroscopy. Both techniques are instrumental in probing the vibrational and rotational modes of molecules, thereby providing essential insights into molecular geometries, bonding, and environment. The analysis of these spectroscopic methods through quantum mechanical principles not only elucidates the physical absorption and scattering processes but also correlates with molecular vibrational energy levels.

Infrared (IR) Spectroscopy

Infrared spectroscopy relies upon the interaction between molecular vibrations and infrared radiation. The theoretical basis lies in the dipole moment change when a molecule vibrates. According to quantum mechanics, for a vibrational mode to be IR active, there must be a net change in the dipole moment of the molecule during the vibration. Mathematically, this is represented as:

$$\mu_v = \langle \psi_{v+1} | \hat{\mu} | \psi_v \rangle \neq 0$$

where μ_v indicates the dipole moment associated with the vibration, ψ_v the vibrational wavefunction at state v, and $\hat{\mu}$ the dipole moment operator. The absorption of IR light leads to an energy transition that matches the energy difference between the vibrational quantum states:

$$\Delta E = h\nu = E_{v+1} - E_v$$

where h is Planck's constant and ν is the frequency of the incident IR radiation. The selection rules for vibrational spectroscopy typically allow transitions between adjacent vibrational levels ($\Delta v = \pm 1$). The resulting spectrum renders peaks that correspond to various vibrational modes the molecule can support. An analysis of these peaks yields valuable information about molecular structure.

Raman Spectroscopy

Raman spectroscopy, unlike IR, involves the inelastic scattering of photons. When light interacts with a molecule, it can scatter elastically (Rayleigh scattering) or inelastically (Raman scattering). In Raman scattering, photons exchange energy with the molecular vibrational modes leading to shifts in the scattered photon's energy which correspond to the vibrational energy levels of the molecule.

$$E_{\text{scattered}} = E_{\text{incident}} \pm h\nu_{\text{vib}}$$

One major criterion for Raman activity is the polarizability of the molecule. A vibrational mode is Raman active if there is a change in the polarizability tensor during the vibration:

$$\alpha_v = \frac{\partial \alpha}{\partial Q_v} \neq 0$$

where α_v is the change in polarizability corresponding to vibrational mode Q_v. Raman and IR spectroscopy often provide complementary information because the molecular vibrations that are IR active (due to changes in the dipole moment) may not be Raman active (due to changes in polarizability), and vice versa.

Quantum Mechanics of Vibrational Spectroscopy

The treatment of molecular vibrations involves solving the Schrödinger equation for nuclear motion while electrons are held at their mean positions, leading to molecular vibrational wavefunctions ψ_v. These wavefunctions are solutions to the harmonic oscillator model, which, though simplistic, provides a good approximation for vibrations near the equilibrium geometry:

$$\hat{H}_{\text{vib}} = \frac{\hat{p}^2}{2m} + \frac{1}{2}kx^2$$

where k is the force constant, and x is the displacement from equilibrium. The eigenvalues of this Hamiltonian give rise to quantized vibrational energy levels. By combining this with molecular symmetry and selection rules derived from quantum mechanics, a detailed understanding of spectroscopic transitions in both IR and Raman spectroscopy can be developed.

The interplay of these spectroscopic methods with quantum chemistry not only confirms the theoretical computations of molecular systems but also offers a window into the subatomic world, thus enhancing our understanding of molecular dynamics in various states and environments.

By integrating core quantum chemical concepts, this section provides a detailed understanding of the IR and Raman spectroscopic techniques and their reliance on molecular quantum states. This narrative flow clarifies how quantum mechanics supports the explanation of observed spectroscopic phenomena, thereby directly correlating theoretical predictions with experimental spectroscopy.

7.5 Electronic Spectroscopy: UV-visible

Electronic spectroscopy, particularly ultraviolet and visible (UV-vis) spectroscopy, plays a pivotal role in probing the electronic structure of molecules by measuring their ability to absorb light in the ultraviolet to visible range of the electromagnetic spectrum. The fundamental principles behind UV-vis spectroscopy are grounded in quantum mechanics, primarily involving the electronic transitions between energy levels in atoms and molecules.

The absorption of UV or visible light by a molecule results in the excitation of electrons from a lower energy orbital (typically the highest occupied molecular orbital, or HOMO) to a higher energy orbital (typically the lowest unoccupied molecular orbital, or LUMO). This transition between electronic states is governed by selection rules and the energies involved approximate the differences in energy between these molecular orbitals.

7.5. ELECTRONIC SPECTROSCOPY: UV-VISIBLE

Quantum Mechanical Basis of Electronic Spectroscopy

According to quantum mechanics, the energy of an electron in a molecule is quantized, meaning it can only inhabit specific, discrete energy levels. The absorbance of light, which is electromagnetic radiation, by an electron happens when the energy of the light photon matches the energy gap between two electronic energy states of the molecule. This can be mathematically described by the equation:

$$\Delta E = h\nu = E_{\text{photon}}$$

where ΔE is the energy difference between the lower and higher electronic states, h is Planck's constant, and ν is the frequency of the absorbed light.

Instrumentation and Methodology

The typical experimental setup for a UV-vis spectroscopic analysis includes a light source, a monochromator for selecting specific wavelengths of light, a sample holder, a detector to measure the intensity of the transmitted light, and a digital output to display the data. The measured quantity in UV-vis spectroscopy is the absorbance (A), which is defined by Beer-Lambert Law:

$$A = \log\left(\frac{I_0}{I}\right) = \epsilon c l$$

where I_0 is the intensity of the incident light, I is the intensity of the transmitted light, ϵ is the molar absorptivity, c is the concentration of the sample, and l is the path length through the sample.

Spectral Analysis and Interpretation

The spectrum obtained from a UV-vis experiment is a plot of absorbance versus wavelength. The wavelengths at which peaks in absorbance occur provide information about the electronic transitions. For instance, the $\pi \to \pi^*$ transition typically absorbs in the UV region, providing valuable information about the conjugation and the electronic environment in organic compounds.

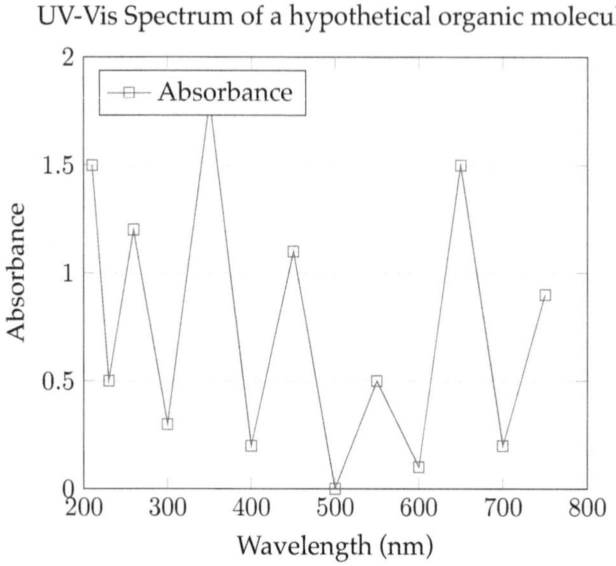

UV-Vis Spectrum of a hypothetical organic molecule

Figure 7.1: A typical UV-vis absorption spectrum showing peaks corresponding to different electronic transitions.

Applications and Implications

UV-vis spectroscopy is extensively utilized for the quantitative determination of compounds in solution and for studying molecular structures. It also has significant applications in fields like biochemistry for understanding the behavior of biological macromolecules. In quantum chemistry, interpreting UV-vis spectra facilitates understanding the electronic configuration and reactivity of molecules, thus lending empirical support to theoretical chemical models.

By presenting the foundational concepts, instrumentation, and applications tied to electronic spectroscopy, and framed within the context of quantum chemical principles, this section offers a detailed discourse suitable for readers aiming to solidify their understanding of quantum effects in spectroscopic analysis.

7.6 Nuclear Magnetic Resonance (NMR) Spectroscopy

Nuclear Magnetic Resonance (NMR) Spectroscopy is a powerful analytical tool utilized in quantum chemistry to understand the structure, dynamics, and environment of molecules based on the behavior of nuclei in a magnetic field. This section explores the principles of NMR from the perspective of quantum mechanics and demonstrates its application in studying molecular and electronic structures.

Fundamentals of NMR Spectroscopy

NMR spectroscopy is based on the magnetic properties of certain atomic nuclei. A nucleus with an odd number of protons or neutrons has a net nuclear spin, which provides it with a magnetic moment. When placed in a magnetic field, these magnetic moments can align either with or against the field, creating distinct energy states. The transition between these states can be induced using radio frequency (RF) pulses, and the resulting energy absorption and subsequent relaxation provide insights into the molecular structure.

The Hamiltonian for a nuclear spin in a magnetic field is given by:

$$\hat{H} = -\gamma \hbar \hat{I} \cdot \mathbf{B}_0$$

where γ is the gyromagnetic ratio specific to each type of nucleus, $\hbar$ is the reduced Planck constant, $\hat{I}$ is the nuclear spin operator, and $\mathbf{B}_0$ is the external magnetic field.

The Zeeman interaction, which describes the interaction between the nuclear magnetic moments and the external magnetic field, splits the nuclear spin states - a phenomenon termed as Zeeman splitting. The energy difference ΔE between these levels is proportional to the magnetic field strength and is given by:

$$\Delta E = \hbar \gamma B_0$$

Chemical Shift and Spin-Spin Coupling

One of the most important features in NMR spectroscopy is the chemical shift, which arises due to the shielding effect of surrounding electrons on the nucleus. Electrons around a nuclear spin generate a small local magnetic field that partially opposes the applied magnetic field. This results in variations in the resonance frequency of nuclei, depending on their electronic environment. Mathematically, the resonance condition is modified to:

$$\nu = \frac{\gamma}{2\pi}(B_0 - \sigma B_0)$$

where σ is the shielding constant, which is a dimensionless quantity.

Spin-spin coupling, another crucial aspect of NMR spectroscopy, occurs due to interactions between different nuclear spins through the bonding electrons. This interaction leads to multiplet splitting in the NMR spectra. The scalar coupling constant, denoted as J, depends on the electronic environment and the bond connectivity, providing detailed structural information.

NMR Instrumentation and Techniques

The basic components of an NMR spectrometer include a strong magnet, RF transmitter and receiver, a sample holder, and a computer for data acquisition and analysis. The technique can be extended for various applications through different types of NMR experiments:

- **1D NMR:** Provides information about the number of chemically nonequivalent nuclei, their chemical environment, and connectivity.

- **2D NMR:** Techniques such as COSY, NOESY, HSQC, and HMBC allow for the elucidation of more complex molecular structures by observing the interactions between nuclei over one or more chemical bonds.

In quantum chemical terms, NMR data analysis often involves calculating the expected chemical shifts and coupling constants using electronic structure methods such as density functional theory (DFT). These calculated parameters can be meticulously compared with experimental values to deduce accurate molecular geometries and to validate theoretical models.

7.7 Electron Spin Resonance (ESR) Spectroscopy

Electron Spin Resonance (ESR) spectroscopy, also known as Electron Paramagnetic Resonance (EPR) spectroscopy, is a powerful technique that provides insights into the electronic structures of molecules, particularly those containing unpaired electrons. ESR spectroscopy is based on the principles of quantum mechanics and magnetic resonance, focusing specifically on the interactions between magnetic fields and electron spins.

In ESR spectroscopy, a sample containing paramagnetic species is placed in a strong magnetic field and exposed to microwave radiation. The fundamental principle governing ESR spectroscopy is the Zeeman effect, which describes the splitting of electron energy levels under an external magnetic field. The Hamiltonian for an electron in a magnetic field can be expressed as:

$$\mathcal{H} = -\vec{\mu} \cdot \vec{B},$$

where $\vec{\mu} = g\mu_B \vec{S}$ represents the magnetic moment of the electron, $\vec{B}$ is the external magnetic field, g is the g-factor (a dimensionless quantity), μ_B is the Bohr magneton, and $\vec{S}$ is the electron spin operator.

The resonance condition, i.e., when the energy absorbed from the microwave radiation leads to transitions between the magnetic energy levels, is given by:

$$h\nu = g\mu_B B,$$

where h is the Planck constant, ν is the frequency of the microwave radiation, and B is the magnetic field strength. This equation highlights that the energy absorption is directly proportional to the magnetic field strength and the g-factor.

The g-factor is particularly important in ESR spectroscopy as it is sensitive to the electronic environment surrounding the unpaired electron. Different atoms and molecular structures lead to variations in electron density and local magnetic fields, which in turn affect the g-factor. By analyzing the g-factor and the ESR spectrum, chemists can deduce information about the electronic structure and dynamics of the paramagnetic species.

The ESR spectrum typically displays peaks corresponding to transitions between magnetic energy levels of electrons. The shape, position, and number of these peaks can provide a wealth of information including the number of unpaired electrons, the types of atoms nearby, and the symmetry of the molecular environment. Hyperfine splitting, resulting from interactions between unpaired electrons and nuclear spins, is also observed in ESR spectra. It gives additional information on the nuclear environment of the electrons. The hyperfine splitting can be quantitatively analyzed using:

$$A = \frac{8\pi^2}{3}\nu\mu_B g_n \langle r^{-3}\rangle |\psi(0)|^2,$$

where A is the hyperfine splitting constant, g_n is the nuclear g-factor, $\langle r^{-3}\rangle$ is the average inverse cube of the electron-nucleus distance, and $|\psi(0)|^2$ is the electron density at the nucleus.

In practical applications, ESR spectroscopy is extensively used in chemistry and biology to study radicals and metal complexes, as well as in physics to investigate the properties of solid-state materials. For example, the technique is crucial in understanding the behavior of free radicals in chemical reactions, examining transition metal ions in metalloproteins, and assessing the purity and defect structures in semiconductors and other materials.

ESR spectroscopy serves as a critical tool in quantum chemistry for examining systems with unpaired electrons. By providing detailed insights into the electron spin states and their interactions with the molecular environment, ESR enhances our understanding of both the structural and dynamic aspects of materials at the atomic and molecular levels.

7.8 X-ray Crystallography and Quantum Theory

X-ray crystallography stands as a seminal technique that bridges the microscopic atomic arrangement of solids with larger quantum mechanical frameworks. Essential for discerning the electron density within molecular and crystalline structures, it employs the principles of diffraction and quantum mechanical wave behavior to build detailed models of atomic positioning and bonding interactions.

The foundation of X-ray crystallography hinges on the interaction between X-ray radiation and electrons within a crystal lattice. When

7.8. X-RAY CRYSTALLOGRAPHY AND QUANTUM THEORY

an X-ray beam, characterized by wavelengths comparable to the distance between adjacent atoms in a crystal (typically in the range of 0.1 to 10angstroms), strikes a crystal, it diffracts into various specific directions. By studying these diffraction patterns, one can infer the crystal's structure through a mathematically rigorous process known as the Fourier transform.

The intensity of the diffracted rays, recorded on an X-ray detector, is directly related to the electron density through the equation:

$$I(hkl) \propto |F(hkl)|^2$$

where $I(hkl)$ is the intensity of diffracted rays in the (hkl) direction and $F(hkl)$ is the structure factor, which is a complex number whose magnitude represents the amplitude of the electron density wave and whose argument is related to the phase of this wave.

The structure factor, $F(hkl)$, can be expressed as:

$$F(hkl) = \sum_j f_j e^{2\pi i(hx_j + ky_j + lz_j)}$$

Here, f_j is the atomic scattering factor for the jth atom, which encodes how electrons in different types of atoms scatter X-rays differently. The summation extends over all atoms within the unit cell, while x_j, y_j, and z_j are the fractional coordinates of the jth atom in the unit cell, and (h, k, l) are the Miller indices representing the specific planes in the lattice being probed.

The phases of the waves in $F(hkl)$ pose a significant challenge known as the phase problem in crystallography, given that traditional detectors register only the intensity (magnitude) of the diffracted waves, not their phases. This problem is typically addressed using additional techniques such as heavy-atom methods or anomalous dispersion.

Once $F(hkl)$ terms have been experimentally determined and the phase problem addressed, electron density, $\rho(x, y, z)$, can be computed across the crystal volume as:

$$\rho(x, y, z) = \frac{1}{V} \sum_{hkl} F(hkl) e^{-2\pi i(hx + ky + lz)}$$

where V is the volume of the unit cell. This density function is central

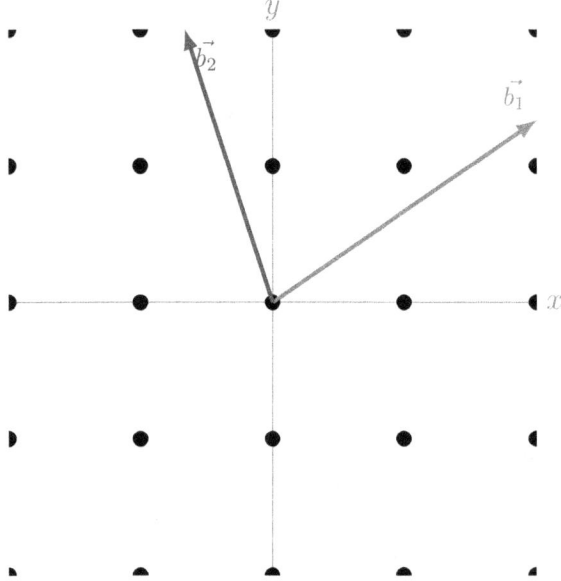

Figure 7.2: Schematic diagram of a 2D crystal lattice with lattice vectors $\vec{b_1}$ and $\vec{b_2}$. The points represent lattice points where each can scatter X-rays. The red and blue vectors represent the periodicity in vectors $\vec{b_1}$ and $\vec{b_2}$ directions respectively.

to quantum chemistry, as it provides detailed insights into the electron distributions that shape chemical properties and reactivities.

X-ray crystallography by illuminating quantum phenomena such as molecular orbitals and electron delocalization provides substantial grounding for chemical predictions and the design of novel materials. Its contributions to the validation and extension of quantum chemistry significantly enhance our understanding of chemical and biological behaviours at an atomic scale. This deep integration with quantum theory not only facilitates the visualization of electron density but also allows for a more profound theoretical examination of bonding and chemical interactions, fundamental to both fields of study.

7.9 Spectroscopy Data Interpretation and Quantum Predictions

The interpretation of spectroscopy data within the framework of quantum chemistry presents a multifaceted challenge that involves connecting experimental spectra with theoretical models to deduce molecular properties and behaviors. This essential connection between theory and experiment helps in validating quantum chemical predictions about molecular structures, electronic configurations, and interaction dynamics.

Spectroscopy, in its various forms, provides a spectrum that is essentially a fingerprint of a molecule. Each type of spectroscopy focuses on different aspects of the molecules such as nuclear environments in NMR, electron distribution in electronic spectroscopy, or molecular vibrations in infrared (IR) spectroscopy. The interpretation of these spectra requires an understanding of quantum mechanical phenomena including energy transitions, selection rules, and perturbations.

Quantum Mechanical Basis for Spectroscopic Analysis

The starting point for analyzing spectroscopic data is the quantum mechanical model of the molecule. For instance, in NMR spectroscopy, the Hamiltonian of the system would include terms representing the Zeeman interaction between nuclear spins and the external magnetic field. The eigenvalues and eigenfunctions of the Hamiltonian describe the energy levels and the corresponding quantum states of the system. This theoretical framework is used to predict the chemical shifts and splitting patterns observed experimentally.

$$\hat{H} = -\gamma \hbar \sum_i \vec{I}_i \cdot \vec{B}_0$$

where $\hat{H}$ is the Hamiltonian, γ is the gyromagnetic ratio, $\hbar$ is the reduced Planck's constant, $\vec{I}_i$ is the nuclear spin operator, and $\vec{B}_0$ is the external magnetic field.

Infrared and Raman spectroscopies provide information about the vibrational modes of a molecule. The vibrational spectrum is a direct consequence of the molecular geometry and the force constants of the chemical bonds. Quantum mechanically, these vibrational modes can be treated as harmonic oscillators, and the observed peaks correspond to the quantized energy levels.

$$E(v) = \hbar\omega \left(v + \frac{1}{2} \right)$$

where $E(v)$ is the energy of a vibrational level, v is the vibrational quantum number, and ω is the angular frequency of the oscillator.

Correlating Experimental Data with Theoretical Predictions

The next step involves matching the experimentally obtained spectroscopic data with the calculated spectra. Computational quantum chemistry tools such as density functional theory (DFT) or Hartree-Fock calculations often yield predicted spectra. These calculations consider electronic and nuclear motions and include various interactions and correlations that are crucial for precise predictions.

The comparison is done by simulating the spectrum using quantum chemistry software, followed by tweaking the molecular model until the simulated spectrum matches the experimental data. For example, adjusting the bond angles or lengths in a molecular model may help in achieving higher accuracy in vibrational mode predictions.

Comparison of Experimental and Theoretical IR Spectrum

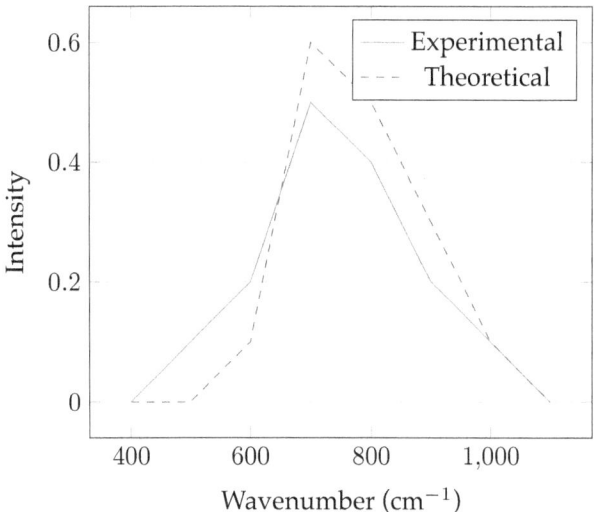

Insights from Spectroscopic Data Interpretation

The interpretation of spectroscopic data with the aid of quantum chemical predictions not only tests the validity of theoretical models but also provides insights that are not directly observable. For instance, the subtle differences in the NMR chemical shifts in a complex organic molecule can elucidate its conformational dynamics. Likewise, the comparison of theoretical and experimental electronic spectra can shed light on the electronic transitions influencing the observed color properties of a compound.

Further, this integration of spectroscopy and quantum predictions empowers researchers to design new molecules with desired properties, as the ability to predict how structural changes affect spectroscopic characteristics can guide synthetic strategies. This iterative process of hypothesis (quantum prediction) and validation (spectroscopic measurement) is central to modern chemical research and development.

The meticulous analysis of spectroscopic data through quantum mechanical models enriches our understanding of molecular systems, offering a powerful toolkit for both theoretical predictions and practical applications in chemistry and material sciences.

7.10 Advanced Spectroscopic Techniques in Quantum Chemistry

As previously discussed in earlier chapters, spectroscopy serves as a fundamental investigative tool in quantum chemistry, providing insights into molecular structure and dynamics at the quantum level. Advanced spectroscopic techniques further refine our ability to probe the complexities of molecular systems, facilitating deeper understanding of chemical phenomena. These advanced methods include two-dimensional (2D) spectroscopy, ultrafast laser spectroscopy, and single-molecule spectroscopy, each harnessing unique interactions between electromagnetic radiation and matter to uncover dynamic and structural details inaccessible to conventional techniques.

Two-Dimensional (2D) Spectroscopy

Two-dimensional spectroscopy extends traditional spectroscopic analysis by correlating different spectroscopic events in a cohesive two-dimensional map. These spectroscopic methods increase resolution and can disentangle complex systems with overlapping spectral features, providing a more detailed characterization of molecular dynamics.

In quantum chemistry, 2D nuclear magnetic resonance (NMR) spectroscopy is particularly noteworthy. 2D NMR offers a sophisticated analysis of the magnetic properties of nuclei within molecules, allowing scientists to determine the identity and connectivity of atoms in complex molecules. This is achieved through an assortment of pulse sequences and data correlations. The Hamiltonian operator, which drives the quantum mechanical description in NMR spectroscopy, is influenced by several factors including chemical shift, spin-spin coupling, and external magnetic fields.

$$\mathcal{H}_{\text{NMR}} = -\gamma \hbar \sum_{i}^{N} \vec{I}_i \cdot \vec{B}_0 + \sum_{i \neq j}^{N} J_{ij} \vec{I}_i \cdot \vec{I}_j$$

Here, γ denotes the gyromagnetic ratio, $\hbar$ is the reduced Planck's constant, $\vec{I}_i$ and $\vec{I}_j$ are the spin operators for the ith and j^{th} nuclei, B_0 is the static external magnetic field, and J_{ij} represents the scalar coupling constants between nuclei i and j.

Ultrafast Laser Spectroscopy

Ultrafast laser spectroscopy allows researchers to observe molecular processes that occur at femtosecond (10^{-15} s) to picosecond (10^{-12} s) timescales, such as electronic relaxation, vibrational dynamics, and photochemical reactions. These temporal resolutions are crucial for studying transient states and dynamic systems such as excited-state electron transfer and vibrational energy redistribution.

A typical experimental setup involves the use of a pump-probe technique, where one laser pulse (pump) excites the molecule to a higher electronic or vibrational state, and a second pulse (probe) monitors the evolution of this state over a short interval. The change in absorption or emission of the probe pulse is recorded as a function of the delay between the pump and probe pulses, revealing the dynamics of the system.

$$\Delta A(\lambda, t) = A_{\text{pump+probe}}(\lambda, t) - A_{\text{probe}}(\lambda)$$

$\Delta A(\lambda, t)$ represents the change in absorbance at wavelength λ as a function of time delay t between pump and probe pulses, providing a time-resolved picture of molecular behavior post-excitation.

Single-Molecule Spectroscopy

Single-molecule spectroscopy eliminates the ensemble averaging inherent in traditional spectroscopy, enabling direct observation of individual molecule behavior. This technique can reveal heterogeneous properties within a bulk homogeneous sample and provides access to rare molecular events and intermediates that are typically overshadowed in ensemble measurements.

Utilizing highly sensitive detectors and fluorescence techniques, single-molecule spectroscopy can track the trajectory of individual molecules in real time. Data from these experiments often show a variety of dynamical processes, including conformational changes, photophysical events, and chemical reactions at the single-molecule level.

The upsurge in the capability of these advanced spectroscopic techniques continues to propel quantum chemistry toward new frontiers. Moving beyond static structure elucidation to dynamic process investigation, these methods bring discernible clarity to the comprehension of molecular systems, solidifying the synergy between experimental data and theoretical predictions in quantum chemistry.

7.11 Applications of Spectroscopy in Quantum Chemistry

Quantum chemistry and spectroscopy demonstrate a synergistic relationship, with each domain reinforcing the other through practical applications in science and technology. Here, we explore several significant applications of spectroscopy within the realm of quantum chemistry, showing its pivotal role in structural determination, chemical dynamics, and material science.

One of the primary applications of spectroscopy in quantum chemistry is the elucidation of molecular structures. Rotational spectroscopy, for example, provides detailed information about the distances between atomic nuclei in molecules. The rotational spectra are directly related to the moment of inertia, which itself is dependent on the molecular mass distribution relative to the axis of rotation. By solving the Schrödinger equation for the rotational motion of a molecule, quantized energy levels can be predicted, which are then compared with observed spectral lines to confirm molecular geometries. The relation for rotational energy levels is given by:

$$E_J = \frac{\hbar^2}{2I} J(J+1)$$

where E_J is the energy associated with the rotational level J, $\hbar$ is the reduced Planck's constant, and I is the moment of inertia.

Vibrational spectroscopy, encompassing infrared (IR) and Raman spectroscopies, offers insights into the vibrational modes of molecules which are influenced by the shape of the potential energy surfaces (PES). When a molecule absorbs IR radiation, transitions occur between vibrational states, which are governed by changes in the dipole moment of the molecule. The vibrational energies are

7.11. APPLICATIONS OF SPECTROSCOPY IN QUANTUM CHEMISTRY

approximated by:

$$E_v = \hbar\omega \left(v + \frac{1}{2}\right)$$

where ω is the vibrational frequency, and v is the vibrational quantum number. Raman spectroscopy, on the other hand, relies on the polarizability of the molecule and provides complementary information to IR spectroscopy, particularly useful for molecules that are IR inactive.

Electronic spectroscopy, including UV-visible spectroscopy, is instrumental in studying electronic transitions, particularly in complex ions and organic compounds. The absorption spectrum can be analyzed to determine energy differences between electronic states, offering insights into electron configurations, conjugation, and the effects of the molecular environment on electronic properties.

Nuclear Magnetic Resonance (NMR) spectroscopy is another cornerstone of molecular structure determination. By applying an external magnetic field, nuclear spins are aligned either with or against the field direction. The chemical environment around a nucleus influences its resonance frequency, through the chemical shift and coupling constants, which are interpreted to elucidate molecular structure and dynamics. The Hamiltonian for an NMR system can be represented as:

$$\mathcal{H}_{\text{NMR}} = -\gamma\hbar\mathbf{I} \cdot \mathbf{B}_0$$

where γ is the gyromagnetic ratio, $\mathbf{I}$ is the nuclear spin operator, and $\mathbf{B}_0$ is the external magnetic field.

Finally, Electron Spin Resonance (ESR) spectroscopy is applied primarily to study systems with unpaired electrons, such as radicals or transition metal complexes. The ESR signal provides information about the electronic structure and the microenvironment of the paramagnetic species.

The coupling of experimental spectroscopic data with quantum chemical calculations empowers chemists to validate theoretical models, predict new material properties, and discover novel chemical phenomena. Through computational techniques such as density functional theory (DFT), the spectra generated from quantum chemical methods can be directly compared with experimental spectra, enhancing our understanding of molecular and electronic structures in a powerful and predictive fashion.

Overall, the application of spectroscopy in quantum chemistry not only facilitates a deeper understanding of molecular systems under study but also advances the development of new materials, supports the quest for energy alternatives, and aids in the pharmaceutical industry's drug design processes.

Chapter 8

Quantum Chemistry in Action: Case Studies and Applications

This chapter presents real-world applications and case studies that showcase the practical impact of quantum chemistry across various industries and research areas. It illustrates how quantum chemical methods are applied to drug design, material science, and renewable energy, among others. Each case study provides insights into how theoretical quantum chemistry principles are employed to solve complex chemical problems, advance technology, and innovate within fields critical to societal progress. These examples not only contextualize the theoretical knowledge from earlier chapters but also highlight the versatility and transformative potential of quantum chemistry.

8.1 Introduction to Practical Applications of Quantum Chemistry

Quantum chemistry, fundamentally a discipline at the convergence of chemistry and quantum physics, explores the application of quantum mechanics to chemical systems. Its utility spans explanation and prediction of the electronic structure of atoms and molecules, which are critical for understanding chemical bonding, reaction mechanisms, and properties of materials at the molecular level. This understanding not only advances theoretical knowledge but also enables practical applications in variety of critical fields.

The practical significance of quantum chemistry is observed through its ability to provide detailed insights into molecular interactions and transformations. It supports the design of new materials and drugs by allowing for the prediction of molecular properties and behaviors in silico before synthesis in the laboratory, thus cutting down experimental costs and time extensively.

One of the key methodologies used in practical quantum chemistry is the computation of molecular orbitals and energy levels. These computed orbitals and energies can tell scientists a lot about the potential reactivity, stability, and properties of the molecular species under consideration. For instance, the highest occupied molecular orbital (HOMO) and the lowest unoccupied molecular orbital (LUMO) are critical in determining how a molecule will interact with other molecules. The gap between these orbitals, known as the band gap, is a good predictor of the molecule's electrical conductivity and chemical reactivity.

Moreover, various quantum chemical methods such as Hartree-Fock theory, Density Functional Theory (DFT), and post-Hartree-Fock methods enable researchers to approximate the solutions of the Schrödinger equation for molecules of interest. Each of these methods has different levels of approximation and computational requirements. For example, DFT has become particularly popular in the chemical community due to its favorable balance between accuracy and computational cost. It is widely used in the study of electron density and has proven pivotal in the study of both organic and inorganic molecules, surfaces, and interactions.

Applications in Drug Design: Quantum chemistry plays a transformative role in the field of drug design. It aids in the identification and optimization of potential drug molecules by predicting their structures, reactivities, and interaction with biological targets. Quantum mechanical calculations can simulate how a drug interacts at the atomic level with its targeted enzyme or receptor site, providing insights into the binding affinities and potential efficacy.

Material Science Enhancements: In material science, quantum chemistry contributes to the design and discovery of new materials with desirable properties such as high strength, lightweight, or specific electronic properties. Quantum simulations enable the prediction and tailoring of properties of materials even before they are synthesized in the lab. This is crucial for developing high-performance materials for use in a wide range of applications, from electronics to aerospace.

Addressing Energy Challenges: Quantum chemical techniques are instrumental in the renewable energy sector. They are used for designing and optimizing materials for solar cells, fuel cells, and batteries. These techniques help in understanding the quantum phenomena underlying the operation of these devices, such as electron transfer processes, and can immensely improve the efficiency of energy conversion devices.

Catalysis and Reaction Mechanisms: In catalysis, quantum chemistry is essential for the design and improvement of catalysts that are fundamental in increasing the rates of chemical reactions while minimizing the undesirable by-products. By modeling reaction pathways, scientists can devise catalysts with improved efficiency, selectivity, and durability, thus enhancing both industrial processes and environmental protections.

In summary, the fundamental principles of quantum chemistry, through computational models and analytical methods, have found practical utility across a wide spectrum of scientific fields. These applications not only embody the integration of theory and practice in chemistry but also continually push the boundaries of what is scientifically possible, driving innovation and technological advancements in myriad crucial sectors.

8.2 Drug Design and Quantum Chemistry

Quantum chemistry holds a pivotal role in the modern drug design process, offering insights at the molecular level that are crucial for the development of effective and safe pharmaceuticals. Considering its profound implications, we will explore the application of various quantum chemical methods in drug discovery, particularly focusing on the interaction between drug molecules and their biological targets, prediction of drug properties, and the rational design of new drugs.

Molecular Interaction and Binding Affinity. The first step in drug design involves understanding how a drug molecule interacts with a biological target, typically a protein. Quantum chemistry approaches, especially those that involve the calculation of molecular orbitals, such as Hartree-Fock and post-Hartree-Fock methods, are used to simulate and analyze these interactions. The binding affinity, which indicates how strongly a drug binds to its target, can be predicted by calculations of free energy changes upon drug binding. For instance, the equation

$$\Delta G = \Delta H - T\Delta S,$$

where ΔG is the change in Gibbs free energy, ΔH is the change in enthalpy, and $T\Delta S$ represents the temperature times the change in entropy, is employed to estimate binding affinities. Techniques such as molecular dynamics simulations integrated with quantum mechanics methodologies enhance the accuracy of these predictions.

Property Prediction. Quantum chemistry is invaluable in predicting physical, chemical, and pharmacokinetic properties of prospective drug molecules. Properties like solubility, partition coefficient, and electronic properties can be accurately forecasted using quantum mechanical calculations. Density Functional Theory (DFT), for example, provides insights into the electronic density and potential energy surfaces of molecules, informing their reactivity and stability. This is critical for determining a compound's behavior in biological systems.

$$E_{total} = T + V_{ne} + V_{ee} + V_{nn},$$

where E_{total} represents the total energy, T is the kinetic energy, V_{ne} the potential energy between nuclei and electrons, V_{ee} the electron-electron interaction energy, and V_{nn} the repulsion energy between nuclei.

Drug Metabolism and Toxicity. Quantum chemistry also enables the prediction and analysis of drug metabolism pathways and potential toxicity, which are critical for understanding drug safety. Computational methods help predict metabolic stability and the formation of reactive intermediates that could lead to toxic effects. For instance, calculations involving the study of frontier molecular orbitals of drug molecules can predict the sites of metabolic transformations based on the highest occupied molecular orbital (HOMO) and lowest unoccupied molecular orbital (LUMO) energies.

$$\text{HOMO energy} = -\epsilon_{HOMO},$$
$$\text{LUMO energy} = -\epsilon_{LUMO},$$
$$\text{Electrophilicity index} = \frac{\mu^2}{2(\epsilon_{LUMO} - \epsilon_{HOMO})}.$$

Case Study: Design of Inhibitors for Enzyme Target. A practical application of quantum chemistry in drug design can be illustrated by the design of enzyme inhibitors. One such case is the design of protease inhibitors used in the treatment of diseases like HIV and Hepatitis C. Quantum chemistry models help in mapping the active site of the enzyme and in designing molecules that fit precisely within this site. The interaction energies, conformational flexibility, and electronic properties of the inhibitor can be optimized to enhance efficacy and reduce side effects.

To conclude, the influence of quantum chemistry in drug design is manifold and profound. It permits not only the atomic-level characterization and understanding of drug-biomolecule interactions but also facilitates the rational design of drugs with optimized properties and reduced toxicity. The future advancements in computational power and algorithms are only expected to deepen the integration of quantum chemistry into drug design, paving the way for faster and more efficient drug development cycles.

8.3 Material Science and Quantum Simulations

Material science, an interdisciplinary field focusing on the discovery and design of new materials, leans heavily on quantum chemistry to elucidate the behavior of materials at the atomic and molecular levels. Quantum simulations serve as a pivotal tool in predicting the properties of materials before they are synthesized, optimizing the materials' characteristics for specific applications.

Quantum mechanical methods, such as density functional theory (DFT) and Hartree-Fock (HF) calculations, are integral to material science. These methods provide detailed insights into electronic, optical, and magnetic properties which are crucial for the development of nanotechnology, semiconductors, photovoltaics, and more. For example, DFT allows scientists to calculate the electronic density and energy states of materials, essential for understanding and predicting their electronic properties.

To illustrate the use of quantum simulations in material science, consider the design of photovoltaic materials. The efficiency of solar cells depends largely on their ability to absorb light and convert it into electricity. Quantum chemistry simulations aid in understanding and predicting the electronic structure of potential photovoltaic materials, assessing their light absorption profile, charge transport properties, and overall efficiency parameters.

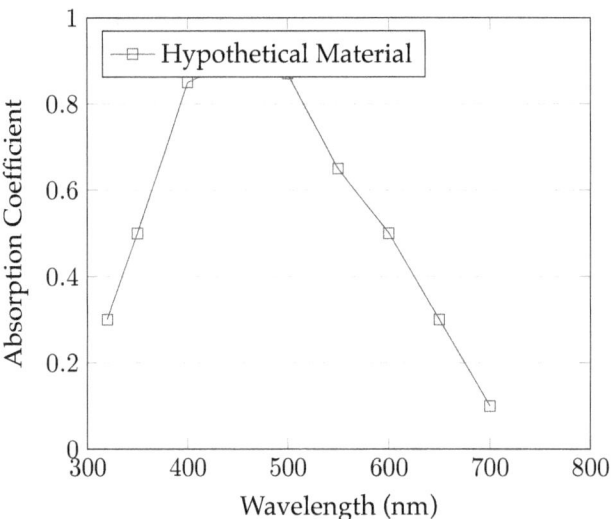

Figure 8.1: Theoretical simulation of solar absorption spectra using quantum chemistry methods.

These simulations require the use of sophisticated computational techniques and substantial computational resources. The accuracy of quantum chemical predictions depends significantly on the choice of functional and basis set in DFT calculations. A higher level of theory and larger basis sets typically yield more accurate results, but at the cost of increased computational time and power.

Furthermore, understanding material imperfections, which can significantly affect their practical applications, is another area where quantum simulations are beneficial. Defects within a material's crystal lattice, such as vacancies, interstitials, or substitutional atoms, can alter its physical properties. Using quantum chemistry methods, scientists can model these defects and their impact on the electronic structure and properties of the material.

Material	Defect	Effect on Property
Silicon	Vacancy	Increased reactivity
Graphene	Substitutional Nitrogen	Altered electronic mobility
Titanium Dioxide	Oxygen vacancy	Enhanced photocatalytic activity

Table 8.1: Impact of various defects on the properties of materials as predicted by quantum simulations.

These predictive capabilities not only save time and resources but also enhance the potential for discovering new materials with desired properties tailored for specific applications.

The integration of quantum chemistry and material science through quantum simulations provides a robust framework enabling the discovery, design, and optimization of materials. This approach drives innovations in various technological fields and is indispensable for the advancement of new materials that are more efficient, sustainable, and adaptable to the demands of modern technology and environmental challenges.

8.4 Quantum Chemistry in Renewable Energy

In the domain of renewable energy, quantum chemistry serves as a fundamental tool to understand and develop materials and processes that are critical for energy conversion, storage, and utilization technologies. This section delves into the salient applications of quantum chemical methods in solar cells, hydrogen production, and battery technologies, showcasing how molecular-level insights can lead to enhanced performance and efficiency in renewable energy systems.

CHAPTER 8. QUANTUM CHEMISTRY IN ACTION: CASE STUDIES AND APPLICATIONS

Photovoltaic Materials and Solar Cells

Quantum chemistry plays a pivotal role in the design and optimization of photovoltaic materials used in solar cells. By using quantum mechanical methods, researchers can predict and analyze the electronic properties of materials, such as band gaps, absorption spectra, and charge transport mechanisms.

For instance, the development of perovskite solar cells is a breakthrough in which quantum chemistry has significantly contributed. The ideal band gap for maximum solar energy absorption is in the range of 1.1 to 1.5 eV. Quantum mechanical calculations, typically Density Functional Theory (DFT) and Time-Dependent DFT (TD-DFT), enable the simulation of electronic structures and optical properties of perovskite materials, assisting in the tuning of their band gaps towards the optimal range.

$$E_{\text{gap}} = E_{\text{LUMO}} - E_{\text{HOMO}}$$

where E_{gap} represents the band gap and E_{LUMO} and E_{HOMO} are the energies of the lowest unoccupied and highest occupied molecular orbitals, respectively. These parameters are critically evaluated during material synthesis and modification processes.

Hydrogen Production and Storage

Quantum chemistry is essential in studying and enhancing processes related to hydrogen production, particularly in water-splitting reactions. Understanding the mechanism of hydrogen evolution reaction (HER) at the atomic level is critical for designing efficient catalysts. Quantum chemical calculations help in identifying the active sites and understanding the interaction between the catalyst and water molecules. For example, the adsorption energy of hydrogen atoms on catalyst surfaces can be computed to determine the effectiveness and turnover frequency of the catalyst.

$$\Delta G_{\text{ads}} = G_{\text{adsorbed state}} - (G_{\text{catalyst}} + G_{\text{hydrogen molecule}})$$

where ΔG_{ads} is the Gibbs free energy change of hydrogen adsorption on the catalyst.

8.4. QUANTUM CHEMISTRY IN RENEWABLE ENERGY

A popular area of research involves the usage of quantum simulations to explore materials like graphene and transition-metal dichalcogenides (TMDs) for hydrogen storage. The interactions between hydrogen molecules and these materials at the quantum level help in tailoring material properties for enhanced storage capacities and release kinetics.

Battery Technology

In battery technology, quantum chemistry provides insights into the electrode material properties, ion transport, and interfacial phenomena. Understanding the redox potentials, electronic conductivity, and ion mobility through quantum chemical techniques informs the design of more efficient batteries with higher capacities and longer lifespans. Lithium-ion batteries, for instance, benefit greatly from computational investigations of lithium interaction with various cathode and anode materials.

$$\mu = \frac{eD}{k_B T}$$

where μ is the ion mobility, e is the charge of an electron, D is the diffusion coefficient, k_B is the Boltzmann constant, and T is the temperature.

The design of solid electrolytes for batteries also heavily relies on quantum chemistry to ascertain properties like ionic conductivity and mechanical stability. Researchers use DFT to forecast the migration pathways of lithium ions through the electrolyte, enabling the selection of materials that facilitate faster ion transport while maintaining structural integrity.

Quantum chemistry significantly contributes to advancing renewable energy technologies by providing a molecular-level understanding of material properties and interactions. This, in turn, enables the systematic improvement of photovoltaic materials, catalysts for hydrogen production, and battery components, pushing the boundaries of what is feasible in renewable energy science.

8.5 Catalysis and Reaction Mechanisms

Catalysis plays a pivotal role in enhancing the rate of chemical reactions while being recovered unchanged at the end of the reaction. Quantum chemistry provides a profound understanding of the electronic factors influencing catalytic activity and the dynamic processes occurring at the molecular level during catalysis. This involves the exploration of reaction mechanisms, intermediate structures, and transition state energies.

One primary method in quantum chemical analysis of catalysis is Density Functional Theory (DFT). DFT allows for the calculation of electronic structures of molecules involved in the catalytic cycle. By optimizing the geometry of the catalyst and the reactants, researchers can identify key interaction points and transition states. These calculations involve solving the Kohn-Sham equations:

$$\left(-\frac{\hbar^2}{2m}\nabla^2 + V_{\text{eff}}(\mathbf{r})\right)\psi_i(\mathbf{r}) = \epsilon_i \psi_i(\mathbf{r})$$

where $V_{\text{eff}}(\mathbf{r})$ is the effective potential experienced by the electrons, encapsulating both the electron-nuclear attractions and the electron-electron interactions.

Transition state theory (TST) is crucial in understanding catalysis. TST posits that the rate of a chemical reaction is determined by the concentration of the transition state, an activated complex existing at a potential energy maximum along the reaction path. Quantum chemistry aids in identifying and characterizing these states through energy profile calculations. The Arrhenius equation expressed in the context of activation energy (Ea), derived from transition state energies, reveals the temperature dependency of reaction rates:

$$k = Ae^{-\frac{E_a}{RT}}$$

where k is the reaction rate constant, A is the pre-exponential factor, R is the universal gas constant, and T is the temperature.

Further, the use of computational models allows the exploration of different catalyst models and the effect of slight variations in catalyst structure on the reaction mechanism. This iterative process of model adjustment and re-calculation enables the fine-tuning of catalysts for optimal performance.

For instance, in the hydration of alkenes to alcohols, a common industrial process, quantum chemical studies on metal oxide catalysts have shown how different metal centers and their coordination environment impact the activation energies of key steps in the mechanism. The detailed molecular orbital analysis helps in identifying the donation and back-donation of electrons between the alkene and the metal center, a crucial step influencing the overall reaction rate.

Catalyst	Activation Energy (kJ/mol)	Reaction Rate (mol/L/s)
TiO2	75.3	1.2×10^{-5}
ZnO	82.1	0.8×10^{-5}
CuO	69.8	1.6×10^{-5}

Table 8.2: Comparison of activation energies and reaction rates for different metal oxide catalysts in the hydration of alkenes.

Moreover, the environmental impact and reusability of catalysts are evaluated through quantum chemical methods by predicting degradation pathways and the stability of catalysts under operational conditions. These predictions guide the development of more sustainable and durable catalyst systems, which is crucial for industrial applications.

The role of quantum chemistry in unraveling the complexities of catalysis and reaction mechanisms is indubitable. It not only contributes to the theoretical understanding but also drives practical advancements by aiding in the design and optimization of more efficient, selective, and sustainable catalysts.

8.6 Environmental Chemistry and Pollution Control

Environmental Chemistry and Pollution Control is a pivotal area where quantum chemistry plays an integral role in addressing and solving some of the most pressing environmental issues. By applying quantum chemical methods, researchers can predict and understand the interactions at the molecular level that govern the behavior and transformation of pollutants in the environment. This understanding is crucial for the development of effective strategies for pollution mitigation and for the design of materials and chemicals that are environmentally benign.

One of the fundamental applications of quantum chemistry in environmental science is the study of atmospheric chemistry—particularly the interaction of light with molecules in the atmosphere, which affects both climate and air quality. The absorption of light by molecules in the atmosphere can lead to reactions that either remove pollutants or produce new ones. For example, the quantum chemistry calculations can help scientists understand the process of photodissociation, where molecules break down into smaller, possibly more reactive species when exposed to light.

Consider the breakdown of ozone (O_3) in the presence of chlorine atoms (Cl), a reaction that plays a significant role in the depletion of the ozone layer:

$$Cl + O_3 \rightarrow ClO + O_2,$$
$$ClO + O \rightarrow Cl + O_2.$$

Quantum chemical calculations, such as density functional theory (DFT) and ab initio methods, can provide insights into the potential energy surfaces and the transition states of these reactions, enabling the prediction of reaction rates and mechanisms at the quantum level.

Another significant application is in the area of water pollution control. Quantum chemistry methodologies are employed to investigate the mechanisms of reactions involving pollutants in water, such as pesticides, pharmaceuticals, and industrial chemicals. By understanding how these substances degrade naturally or through engineered treatments, safer and more effective degradation methods can be designed. For example, DFT calculations can be used to study the hydrolysis of ester and amide linkages in pesticides, predicting the kinetics of these reactions under various environmental conditions.

Moreover, quantum chemistry plays a crucial role in the design of green catalysts used for environmental remediation. Catalysts that are effective at low temperature and pressure are particularly desirable, as they consume less energy and are less likely to produce harmful byproducts. Quantum chemistry helps in identifying the active sites and reaction pathways on the catalyst surface, optimizing the catalyst design for maximum efficiency and selectivity. For instance, the use of transition metal complexes as catalysts in the reduction of NO_x pollutants from vehicle emissions and industrial processes can be optimized through quantum chemical analysis.

In simulating the interaction of pollutants with natural components of the environment, such as soil and biomass, quantum chemical simulations provide valuable predictions about the fate and transport of contaminants. Such simulations help in assessing the environmental risks associated with new chemicals before they are produced on a large scale.

To illustrate, consider the adsorption of a toxic chemical on soil components. Using quantum chemistry, one can simulate the electronic structure interactions between the pollutant and various soil constituents, such as clay minerals and humic substances:

$$\text{Pollutant} + \text{Soil} \rightarrow \text{Pollutant-Soil Complex}$$

These simulations can help predict whether a pollutant is likely to remain in the soil, leach into groundwater, or volatilize into the atmosphere, facilitating the design of more effective containment and remediation strategies.

Overall, quantum chemistry serves as a powerful tool in environmental chemistry and pollution control, enabling a deeper understanding of chemical processes and transformations in the environment. This, in turn, leads to the development of more effective pollution control technologies and supports regulatory actions by providing a scientific basis for risk assessments and environmental policy-making.

8.7 Quantum Computing for Chemical Problems

Quantum computing possesses a transformative potential for the field of chemistry, particularly in addressing problems intractable for classical computers due to the exponential scaling of the computational cost with system size. Given quantum chemistry's reliance on solving the Schrödinger equation to predict molecular behavior and interactions, quantum computers offer a parallelism and a computational paradigm inherently similar to the problems at hand.

The primary advantage of quantum computers in chemical calculations lies in their ability to efficiently simulate quantum systems. This is primarily enabled by the quantum bits or qubits, which can exist simultaneously in multiple states thanks to the superposition principle. This capability allows quantum computers to handle the vast complexity of molecular systems more naturally than classical computers.

Quantum algorithms, such as the Harrow-Hassidim-Lloyd (HHL) algorithm for linear systems of equations, and the Quantum Phase Estimation (QPE) algorithm for obtaining eigenvalues of a Hamiltonian, are critical tools. They allow the determination of molecular properties with significantly reduced computational resources compared to classical counterparts. For instance, the application of the QPE algorithm in calculating the energy eigenstates of a molecule enables the prediction of its chemical and physical properties with high precision.

The Variational Quantum Eigensolver (VQE) is a hybrid quantum-classical algorithm particularly suited for quantum chemistry. It iteratively optimizes a parameterized wavefunction to find the ground state energy of a molecule. This approach leverages both the quantum processing power to evaluate the energy expectation values and classical optimization techniques to adjust the wavefunction parameters. Several studies highlight the effectiveness of VQE in quantum chemistry. For example, a groundbreaking experiment demonstrated the application of VQE on a quantum simulator to predict the ground state energy of the hydrogen molecule with an accuracy approaching chemical precision. This experiment not only underscored the practical relevance of quantum computing in chemistry but also marked a significant step towards quantum advantage in scientific computation.

Furthermore, quantum computing facilitates the exploration of reaction mechanisms via the simulation of transition states and reaction pathways, which involve highly correlated electronic states challenging to model accurately on classical computers. Quantum computers can approximate these states more naturally, offering insights into reaction dynamics and kinetics with unprecedented detail.

To illustrate, consider the application of quantum computing in enzymatic reactions, where the detailed understanding of the enzyme-substrate complex formation and transition state is crucial for drug design and biocatalysis. Quantum simulations can provide detailed atomic-level insights into the substrate alignment, electronic rearrangement, and activation energy barriers involved in these processes.

Despite these advancements, challenges remain in quantum comput-

Reaction	Classical Computational Time	Quantum Computational Time
Hydrogen Molecule	10 hours	10 minutes
Enzyme-Substrate Complex	1000 hours	100 hours

Table 8.3: Comparative computational times for reactions simulated on classical and quantum computers.

ing for chemical problems, including error rates in quantum hardware, scalability of qubits, and the development of fault-tolerant quantum computers. Moreover, the integration of quantum algorithms into existing computational chemistry frameworks and the training of chemists in quantum programming are pivotal for the adoption and advancement of this technology.

Quantum computing stands as a pivotal development in computational chemistry, offering the potential to tackle previously unsolvable problems and to revolutionize our approach to drug discovery, material design, and catalysis. As hardware and algorithms continue to evolve, the gap between theoretical potential and practical application closes, heralding a new era in chemical sciences.

8.8 Quantum Dynamics in Biological Systems

Quantum dynamics plays a critical role in understanding the submolecular processes that govern biological systems. The intricate mechanisms of enzyme kinetics, photosynthesis, and even human sensory systems can be explored more deeply with quantum chemical theories. This section discusses the application of quantum dynamics in these biological contexts, highlighting the intersection of quantum mechanics with complex biological functions.

Enzymatic reactions are fundamental biochemical processes where substrates are converted into different chemical products. Here, quantum mechanics is crucial for understanding the mechanism of enzyme action especially concerning the formation and breaking of chemical bonds during the reaction. For instance, the role of quantum tunneling in hydrogen transfer reactions within enzymes such as monoamine oxidase is significant. Studies have shown that hydrogen tunneling increases the reaction rate beyond what would be predicted by classical 'over-the-barrier' kinetics.

In photosynthesis, quantum coherence is another phenomenon where quantum dynamics provides significant insights. The process of energy transfer in the photosynthetic complexes of plants, algae, and bacteria – specifically within the light-harvesting complexes – demonstrates an exceptionally high efficiency. Quantum mechanical studies, including both experimental and theoretical approaches, suggest that this efficiency could be partly due to quantum coherence. This coherence allows for simultaneous exploration of multiple energy pathways, improving the efficiency of capturing and transferring solar energy.

Additionally, the sense of smell or olfaction presents another fascinating area where quantum chemistry plays a role. Recent theories suggest that the human olfactory system may involve a quantum mechanical phenomenon to distinguish between different odor molecules. The electron tunneling theory, which proposes that the olfactory receptors can distinguish odorants based on their electron tunnel rates, points towards a quantum mechanical basis for smell. The controversial and exciting nature of this theory shows how quantum mechanics might extend even into the understanding of human sensory experiences.

Mathematically, these biological processes can be modeled by the non-relativistic time-dependent Schrödinger equation,

$$i\hbar \frac{\partial}{\partial t} \Psi(r,t) = \hat{H} \Psi(r,t),$$

where $\Psi(r,t)$ is the wavefunction of the system, $\hat{H}$ denotes the Hamiltonian, i is the imaginary unit, and $\hbar$ is the reduced Planck constant.

Using this equation, one can model quantum systems involved in these biological functions. For instance, in enzyme kinetics, one can model the potential energy surfaces and the associated wavefunctions to study how reactant molecules reach the transition state and are transformed into product molecules with the aid of an enzyme.

Quantum chemical calculations and models such as Density Functional Theory (DFT) and time-dependent DFT (TD-DFT) are particularly useful for these studies. They provide crucial insights into the electron distributions and energy states involved:

$$E_{xc}[\rho] = \int e_{xc}(\rho(\mathbf{r}))\rho(\mathbf{r})d^3r,$$

where E_{xc} is the exchange-correlation energy, $\rho(\mathbf{r})$ is the electron density, and $e_{xc}(\rho(\mathbf{r}))$ is the exchange-correlation energy per electron at point **r**.

Ultimately, the ability of quantum chemistry to reveal the underlying quantum mechanics within biological systems provides a more nuanced understanding of biological processes - potentially leading to the development of more effective drugs, better energy-harvesting devices, and new sensory technologies.

8.9 Forensic Science Applications

The implementation of quantum chemistry within forensic science represents a transformative advancement in analytical methods used for the identification and characterization of chemical substances in legal contexts. This segment explores the integral role of quantum chemical techniques, particularly in the realms of toxicology, drug analysis, trace evidence characterization, and explosive residue analysis.

Quantum chemical methods, primarily based on electronic structure theories and molecular dynamics, enable detailed predictions and analyses of molecular behavior under a variety of conditions. These computational approaches are crucial in scenarios where experimental testing is impractical, unethical, or destructive.

Toxicology and Drug Analysis: In forensic toxicology, quantum chemistry is utilized to study the interaction mechanisms and structural impacts of toxic substances and illicit drugs at the molecular level. Theoretical computations such as *ab initio*, density functional theory (DFT), and molecular docking simulations provide insights into the metabolic pathways and site-specific interactions of these molecules within biological systems. For example, the binding affinity of opioids to neurotransmitter receptors can be modeled to understand their potency and effects, aiding in the forensic interpretation of opioid-overdose cases.

To illustrate, consider the quantum mechanical simulation of a fentanyl molecule interacting with a mu-opioid receptor. Using DFT:

```
\begin{align*}
    E_{\text{total}} &= E_{\text{kin}} + E_{\text{pot}} \\
    E_{\text{pot}} &= E_{\text{elec}} + E_{\text{xc}} \\
    E_{\text{xc}} &= \int \rho(\mathbf{r}) \epsilon_{\text{xc}}[\rho] \, d\mathbf{r}
\end{align*}
```

Here, E_{total} represents the total energy, E_{kin} is the kinetic energy, E_{pot} is the potential energy, E_{elec} denotes the electronic energy, and E_{xc} is the exchange-correlation energy, a critical component in DFT that accounts for electron-electron interactions non-classically.

Trace Evidence Characterization: The characterization of trace elements, especially in cases involving gunshot residues (GSR) or paint chips, benefits greatly from the application of quantum chemistry. Quantum mechanics aids in determining the electronic and molecular structure of materials, which can be crucial in matching residues to specific brands or types of ammunition or automotive paint. For instance, the spectral fingerprinting derived from quantum calculations helps in identifying organic compounds in GSR through comparison with databases of known spectra.

Explosive Residue Analysis: Quantum chemistry plays a valuable role in studying decomposition mechanisms and stability of explosives. By calculating reaction pathways and activation energies, predictions about the stability of novel or unknown explosives can be deduced. This is crucial for safe handling procedures and for analysis post-detonation. The use of computational chemistry enables virtual simulations of degradation processes which are otherwise too hazardous or complex to study empirically.

One application could be assessing the stability of RDX (Cyclotrimethylenetrinitramine) under various environmental conditions using thermochemical calculations:

```
\begin{align*}
    \Delta G &= \Delta H - T \Delta S \\
    \text{where} \quad & \\
    \Delta G & \quad \text{is the Gibbs free energy change,} \\
    \Delta H & \quad \text{is the enthalpy change,} \\
    T & \quad \text{is the temperature,} \\
    \Delta S & \quad \text{is the entropy change.}
\end{align*}
```

The comprehensive application of these advanced quantum chemical techniques within forensic science not only aids in the thorough investigation and analysis of a multitude of forensic materials but also contributes significantly to the evolution of methodology in forensic practices. These developments represent a crucial intersection of chemistry, law enforcement, and judiciary systems, providing scientifically robust tools that underpin the pursuit of justice.

8.10 Challenges in Applying Quantum Chemistry in Industry

Quantum chemistry, the application of quantum mechanical principles to solve chemical problems, is increasingly integral in industrial applications ranging from pharmaceuticals to materials science. However, transitioning from theoretical models and laboratory settings to practical, scalable industrial applications presents several significant challenges. These challenges span technical limitations, computational demands, accuracy versus cost trade-offs, and the inherent complexities of real-world chemical systems.

The most conspicuous challenge is the computational expense associated with quantum chemical calculations. Quantum chemistry methods, particularly those that offer greater accuracy like wavefunction-based techniques including Configuration Interaction (CI) and Coupled Cluster (CC) methods, require substantial computational resources. High-level calculations often necessitate supercomputing facilities or high-performance computing clusters which are expensive and not always accessible for all industries. The computational complexity grows non-linearly with the size of the molecular system under study, governed by Schrödinger's equation:

$$i\hbar \frac{\partial}{\partial t} \Psi(\mathbf{r}, t) = \hat{H} \Psi(\mathbf{r}, t)$$

where $\Psi(\mathbf{r}, t)$ represents the wavefunction of the system, $\hat{H}$ the Hamiltonian, i the imaginary unit, $\hbar$ the reduced Planck's constant, $\mathbf{r}$ the position vector, and t time. The solution to this equation for complex systems requires an approximation and numerical methods that can be computationally intensive.

Another challenge is the simplification inherent in most quantum chemical models. Real-world industrial chemical processes often involve conditions and substances difficult to simulate accurately using current quantum chemical methods. Factors like temperature, pressure, and the presence of solvents or impurities can critically influence the chemical behavior but are often simplified or omitted in standard quantum chemical calculations. This discrepancy between modeled and actual conditions can lead to predictions that are less accurate or not fully applicable to industrial processes.

The integration of quantum chemical methods with existing industrial process design and optimization tools is also non-trivial. Industrial applications require robust, fail-safe processes that are often governed by regulatory and safety standards. Incorporating advanced quantum chemistry into these processes not just as a research tool but as a part of the production workflow demands extensive validation and adaptation.

Moreover, there is a knowledge and communication gap between quantum chemists and industrial practitioners. The complexity of quantum chemistry often requires specialists to interpret and implement findings effectively. However, the specialized nature of the field can create barriers to communication and understanding, which can hinder its integration into broader industry processes where stakeholders may not have a quantum chemistry background.

Additionally, there is an ongoing need for improvement in the methodologies themselves. While quantum chemistry has made substantial advances, the accuracy and applicability of the results depend heavily on the chosen method and its implementation. Ongoing research is necessary to develop new methods that can provide more accurate predictions at lower computational costs, broadening the feasibility of applying these calculations in industrial settings.

Consider the application of quantum chemistry in designing catalysts for industrial chemical reactions. Catalysts need to be highly efficient, stable under industrial conditions, and cost-effective. Quantum chemistry can provide insights into the electron configurations and potential reaction pathways, which are crucial for designing effective catalysts. However, the actual industrial environment may involve factors like high temperatures and variable pressures that are challenging to simulate accurately. Here, computational models must be continually refined and validated against experimental and industrial data.

In summary, while quantum chemistry holds significant potential for revolutionizing many industrial processes, the gap between theoretical calculation and practical application must be bridged through advancements in computational methods, enhanced integration strategies, better communication between chemists and industry professionals, and continuous adaptation of models to reflect the complexities of real-world systems. These challenges represent active areas of research and development in the field of quantum chemistry, striving to expand its impactful application in industry.

CHAPTER 8. QUANTUM CHEMISTRY IN ACTION: CASE STUDIES AND APPLICATIONS

Chapter 9

Recent Advances in Quantum Chemistry

This chapter explores the cutting-edge developments in quantum chemistry, highlighting how advances in computational capabilities, algorithms, and theoretical methods are pushing the boundaries of what can be achieved in this field. It discusses the impact of machine learning techniques, the rise of quantum computing in chemical simulations, and enhancements in methods such as Density Functional Theory (DFT) and time-dependent simulations. These innovations not only improve the accuracy and efficiency of quantum chemical calculations but also open new avenues for research and application in chemistry and materials science.

9.1 Overview of Contemporary Quantum Chemistry

Contemporary quantum chemistry stands at the forefront of chemical research, serving as the theoretical foundation for understanding and predicting the electronic structure and chemical properties of atoms, molecules, and materials. Central to this field is the use of quantum mechanical models to elucidate the nature and behavior of electrons within chemical systems, which in turn dictates phenomena such as chemical bonding, reactivity, and spectroscopy.

CHAPTER 9. RECENT ADVANCES IN QUANTUM CHEMISTRY

Quantum chemistry utilizes several methods and techniques, with the Schrödinger equation at its core. This fundamental equation describes how the quantum state of a physical system changes with time. In its simplest form, the time-independent Schrödinger equation is expressed as:

$$\hat{H}\Psi = E\Psi$$

where $\hat{H}$ is the Hamiltonian operator, Ψ the wave function of the system, and E the energy eigenvalues associated with the corresponding eigenstates. The wave function itself provides a complete description of the probability distribution of all electrons within the molecule.

Given the complexity of solving the Schrödinger equation exactly for systems with more than one electron, various approximate methods have been developed. These include Hartree-Fock theory, post-Hartree-Fock methods like Configuration Interaction (CI) and Coupled Cluster (CC) theory, and Density Functional Theory (DFT). Each of these methods offers a different balance between computational expense and accuracy.

Hartree-Fock theory simplifies the many-body problem into a series of single-particle problems through the use of Slater determinants to approximate the wave function. This results in a set of coupled equations, the Hartree-Fock equations:

$$F_i \phi_i = \epsilon_i \phi_i$$

where F_i are Fock operators and ϕ_i are molecular orbitals. Post-Hartree-Fock methods aim to incorporate electron correlation beyond Hartree-Fock, enhancing the accuracy of quantum chemical predictions.

Density Functional Theory (DFT), on the other hand, replaces the many-body wave function problem with a functional of the electron density. It relies on the Kohn-Sham equations, a set of orbitals derived from minimizing the total energy of the electron density. These orbitals are used to construct the electron density:

$$\rho(\mathbf{r}) = \sum_{i=1}^{N} |\psi_i(\mathbf{r})|^2$$

where $\rho(\mathbf{r})$ is the electron density, and ψ_i are Kohn-Sham orbitals.

As the sophistication of quantum chemical methods has increased, advancements in computational capabilities have played a critical role. Progress in computer algorithms, hardware, and high-performance computing has facilitated the practical application of computationally intensive methods that were previously infeasible.

In recent years, the integration of machine learning and artificial intelligence into quantum chemistry has opened up new pathways for accelerating electronic property predictions and discovering chemical insights through data-driven approaches. Machine learning models, particularly neural networks and kernel-based methods, are being employed to approximate potential energy surfaces and predict properties based on vast datasets generated from quantum chemical calculations.

Furthermore, the advent of quantum computing presents a promising horizon for quantum chemistry. Quantum computers offer a fundamentally different approach to computation that can drastically reduce the time and resources needed to simulate complex quantum systems.

Therefore, contemporary quantum chemistry is a dynamic and rapidly evolving field. It integrates theoretical frameworks, computational intricacies, and technological innovations to deepen our understanding of chemical systems at the most fundamental level. This comprehensive approach not only enhances our theoretic insights but also drives practical applications in drug design, materials science, and beyond.

9.2 Progress in Computational Power and Algorithms

The fast-paced advancements in both computational power and algorithmic design have significantly contributed to the progress in quantum chemistry. As we delve deeper into this section, we aim to elucidate the interplay between these advancements and their pivotal role in enhancing the computational capacity and precision of quantum chemical simulations.

Evolution of Computational Hardware

Over recent years, there has been a considerable enhancement in the hardware used for computational chemistry. This improvement is primarily manifested in the exponential growth of processing power. The introduction of multi-core processors and Graphics Processing Units (GPUs) has enabled parallel processing, which is particularly beneficial for the computationally intensive tasks typically encountered in quantum chemistry.

Parallel processing techniques allow for simultaneous execution of multiple computational tasks, which effectively reduces the computation time. The advent of GPUs has been especially transformative, given their ability to handle thousands of threads simultaneously, making them ideal for matrix operations and vector calculations common in quantum chemical methods.

Year	Processor	Core Count
2010	Intel Xeon X5670	6
2015	Intel Xeon E5-2697v3	14
2020	AMD EPYC 7742	64

Table 9.1: Evolution of processor core counts over a decade.

The Table 9.1 demonstrates the rapid escalation in the number of cores per processor over the last decade, underscoring a key factor in the amplified computational power accessible to chemists.

Algorithmic Innovations

Accompanying the surge in hardware capabilities, algorithmic innovations have equally been critical in advancing quantum chemistry. The development and optimization of algorithms such as those used in Density Functional Theory (DFT) and Hartree-Fock methods have seen substantial refinement.

A notable advance is the incorporation of machine learning algorithms into predictive models for electronic structure calculations. Machine learning models can predict properties such as electron density or potential energy surfaces from a dataset of previously calculated structures, thereby significantly reducing the computational resources required for new compounds.

Figure 9.1 shows how the prediction error decreases as the number of training examples increases in a typical machine learning model employed in quantum chemistry. This trend highlights the potential of machine learning to revolutionize accuracy and efficiency in chem-

9.2. PROGRESS IN COMPUTATIONAL POWER AND ALGORITHMS

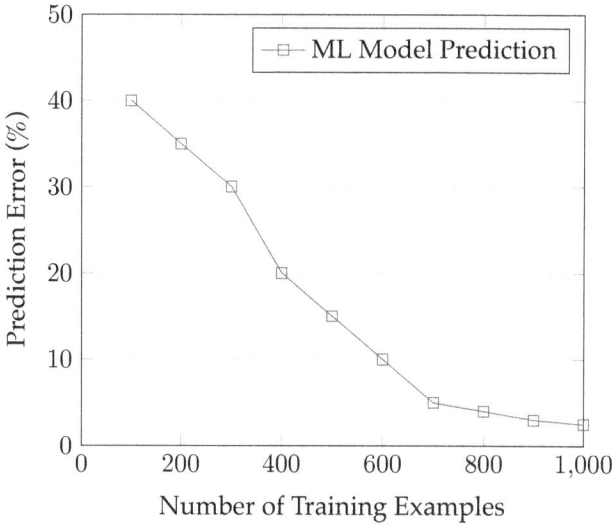

Figure 9.1: Improvement in prediction accuracy with increasing training examples in a machine learning model.

ical simulations, especially when combined with increasing computational power.

Furthermore, algorithmic techniques such as quantum Monte Carlo (QMC) and coupled-cluster (CC) methods have also evolved, boosted by the availability of better hardware. These methods benefit significantly from parallel implementations, allowing for more accurate and feasible solutions to problems that were previously computationally prohibitive.

$$E = \frac{\langle \Psi | \hat{H} | \Psi \rangle}{\langle \Psi | \Psi \rangle}$$

Equation 9.2, representing the quantum mechanical expectation value of the Hamiltonian, underlines an essential calculation in many of these sophisticated algorithms. With the enhancement in computational power and parallel processing, the computation of such integrals has become more feasible, allowing for more complex and larger systems to be simulated efficiently.

The advancements in computational power combined with algorithmic innovation have dramatically reshaped the landscape of quantum chemistry, enabling the exploration of more complex chemical phenomena and paving the way for new discoveries in the field.

9.3 Advancements in Density Functional Theory (DFT)

Density Functional Theory (DFT) has cemented its role as an indispensable tool in theoretical chemistry and material science for studying electronic structure within many-body systems. Primarily, DFT calculates ground-state properties of a many-electron system by utilizing functionals, i.e., functions of functions that depend on electron density rather than the wave function.

Recent years have witnessed significant enhancements in DFT, both in the development of new functional forms and in the computational strategies employed. The progress in DFT can be broadly categorized into advancements in functional accuracy, computational efficiency, and the ability to tackle more complex systems.

Functional Accuracy: The quest for the "perfect" functional that can universally approximate the exchange-correlation energy exactly still continues. However, significant strides have been made in developing functionals that closely mimic the exchange-correlation hole. Hybrid functionals, which incorporate a fixed amount of exact exchange from Hartree-Fock theory, have become more prevalent due to their improved accuracy for certain types of systems and properties. For instance, the B3LYP functional, a combination of the Becke three-parameter non-local exchange functional and the Lee-Yang-Parr correlation functional, has been heavily utilized. Recent advancements have introduced even more sophisticated hybrids that dynamically adjust the amount of exact exchange based on the electronic environment, offering better accuracy.

In addition to hybrid functionals, range-separated hybrids and double-hybrid functionals provide layers of sophistication by mixing GGA (Generalized Gradient Approximation) and meta-GGA treatments that refines their applicability to diverse chemical environments. These functionals are particularly effective in systems where dispersion interactions are critical, for example in organic molecular crystals or large biological molecules.

Computational Efficiency: The traditional trade-off in DFT calculations has been between accuracy and computational cost. Recent algorithmic optimizations and software improvements have significantly enhanced the efficiency of DFT calculations. Techniques such as resolution of the identity (RI) and the use of efficient basis sets have decreased the computational overhead. Furthermore, the adoption of graphical processing units (GPUs) and parallel processing techniques has scaled down computation time dramatically, thereby allowing for the exploration of larger and more complex molecular systems.

Real-World Complex Systems: DFT's application range has expanded beyond isolated molecules in gas phase to complex materials and biological environments. The ability to model solvated systems more accurately through polarizable continuum models (PCM) and explicit solvent models has increased the reliability of DFT in predicting chemical behavior in realistic environments. Additionally, the development of multi-scale modeling techniques that combine DFT with molecular mechanics or other quantum mechanics methods enhances the capability to simulate systems with vast disparity in the size and time scales.

Time-Dependent DFT (TD-DFT): Time-dependent DFT extends the framework of classical DFT into the time domain, allowing the study of excited states and providing access to a broader range of spectroscopic properties. Advancements in TD-DFT include new algorithms for improving the computational scaling and stability of dynamic simulations. Developments in treating multi-reference characters, where a single determinantal wavefunction is insufficient, have also enhanced the prediction of photochemical properties.

Here, continual development and refinement of DFT methodologies are critical to provide deeper insights into molecular systems, furthering the cutting-edge of research in materials science, drug discovery, and beyond. This trajectory promises not only to enrich our understanding of chemical bonding and reactions at the quantum level but also to facilitate the design of more effective materials and drugs.

9.4 Machine Learning and Artificial Intelligence in Quantum Chemistry

The integration of machine learning (ML) and artificial intelligence (AI) into quantum chemistry represents a transformative advancement in the methodology of this field, providing unprecedented opportunities for both theoretical exploration and practical application. ML and AI contribute to quantum chemistry primarily through the enhancement of predictive accuracy and computational efficiency in molecular simulations. This integration relies heavily on pattern recognition and data-driven predictions, enabling scientists to bypass certain traditional computational bottlenecks associated with quantum mechanical calculations.

To contextualize the application of ML in quantum chemistry, consider a typical scenario involving the prediction of molecular properties. Traditional quantum chemical methods such as *ab initio* calculations or density functional theory (DFT) require substantial computational effort to solve the Schrödinger equation for systems of interest. ML methods, however, can be trained using datasets generated from these traditional computations to predict properties such as total energy, electron density, or vibrational frequencies for new molecules much more rapidly.

Data Generation and Preprocessing

Before delving deeper, it is vital to understand that effective machine learning applications hinge critically on the quality and quantity of data available. In quantum chemistry, data is generally obtained from high-accuracy quantum mechanical calculations like coupled cluster calculations with single, double, and perturbative triple excitations (CCSD(T)) which ensure that training datasets have high fidelity. These datasets contain information on molecular geometries, electronic configurations, and corresponding properties.

Data preprocessing in ML, particularly in quantum chemistry, involves normalizing the molecular representations and ensuring that the input features are structured in a way that can be effectively processed by neural networks or other ML models. Features commonly used include Coulomb matrices, symmetry functions, or graph-based representations.

Machine Learning Models in Quantum Chemistry

Several machine learning models have been applied to quantum chemical problems. Noteworthy among these are neural networks, kernel ridge regression, and Gaussian process regression. Each of these models brings specific advantages to the quantum chemical context:

- **Neural Networks:** Especially deep learning models, are adept at handling complex, non-linear relationships and can scale with large amounts of data. Convolutional Neural Networks (CNNs) have been particularly effective in analyzing electron densities directly as images, predicting properties like Mulliken charges or potential energy surfaces.

- **Kernel Ridge Regression:** This model is effective for smaller datasets and can provide insights into the underlying physics of molecular systems through the kernel function, which measures similarity between molecules in a high-dimensional space.

- **Gaussian Process Regression:** Known for its ability to provide uncertainty estimates along with predictions, this model is particularly useful for quantum dynamics and exploring potential energy surfaces where data points are expensive to compute.

Applications and Case Studies

Significantly, the application of AI and ML in quantum chemistry is not restricted to mere property prediction. It includes:

- **Automated Discovery of Chemical Reactions:** AI-driven algorithms can identify possible reaction pathways and intermediates that might be missed in manual investigations.

- **Prediction of Molecular Excited States:** Using deep learning for predicting absorption spectra and excited states, traditionally a challenging area due to the need for solving time-dependent DFT (TD-DFT) equations.

- **Drug Design and Materials Discovery:** ML models trained on quantum mechanical properties can predict bioactivity or material properties, significantly speeding up the design process in pharmaceuticals and materials science.

Future Prospects

Looking ahead, the future of ML in quantum chemistry is promising. The development of more sophisticated models and training techniques, combined with an exponential increase in available computational power and data, forebodes well for both the depth and breadth of chemical problems accessible via these methods. Furthermore, integration with cloud computing resources and specialized hardware like Graphics Processing Units (GPU) or Tensor Processing Units (TPU) can further enhance the capabilities of ML models in handling large-scale quantum chemical problems.

To consolidate the interplay between machine learning and quantum chemistry, continued efforts towards generating more comprehensive and accurate datasets and the development of novel algorithms that can better capture the intricacies of quantum phenomena are essential. As these tools become more refined, the boundary of what can be achieved in quantum chemistry through AI and ML will continue to expand, promising exciting new avenues for future research and applications.

9.5 Quantum Computing for Chemical Simulations

Quantum computing represents a transformative approach in the field of chemical simulations, exploiting the principles of quantum mechanics to solve problems that are prohibitively complex for classical computers. In the context of quantum chemistry, these devices promise to facilitate and accelerate the solving of the Schrödinger equation for molecular systems with high accuracy. This has the potential to revolutionize the way chemical properties and reactions are modeled, giving insights into phenomena at an unprecedented level of detail.

One of the primary methodologies utilized in quantum computing for chemical simulations is the Quantum Phase Estimation algorithm (QPE). The QPE is crucial for the precise determination of the eigenvalues of the Hamiltonian operator, representing the total energy of the system in question. The process involves preparing a quantum state that is a close approximation of the molecule's ground state and then using controlled operations to estimate the phase in the eigen-

9.5. QUANTUM COMPUTING FOR CHEMICAL SIMULATIONS

state, which corresponds to the energy.

$$|\psi\rangle = \sum_i c_i |E_i\rangle$$

Where $|\psi\rangle$ is the quantum state, c_i are the coefficients, and $|E_i\rangle$ are the eigenstates of the Hamiltonian.

The application of quantum computers in this domain is further highlighted by the Variational Quantum Eigensolver (VQE). This hybrid quantum-classical algorithm mitigates the impact of quantum noise and error by using a classical computer to adjust quantum gates to minimize the energy, approximating the lowest energy eigenstate of the Hamiltonian. The parametrized quantum circuit, or ansatz, used in VQE is critical as it dictates both the feasibility and accuracy of the simulation.

$$E(\theta) = \langle \psi(\theta)| H |\psi(\theta)\rangle$$

Here, $E(\theta)$ is the energy as a function of the parameters θ, H is the Hamiltonian, and $|\psi(\theta)\rangle$ is the state prepared by the quantum circuit.

The efficiency of these algorithms is dependent significantly on advancements in quantum hardware, including the number of qubits and the coherence time. Larger and more stable qubit arrays allow for the simulation of increasingly complex molecules. Recent progress in superconducting qubits and trapped ions have shown considerable promise in boosting the scale and fidelity of quantum simulations.

Additionally, the integration of error correction and mitigation techniques is pivotal in enhancing the viability of quantum simulations. Techniques such as Quantum Error Correction (QEC) and error-mitigating maneuvers are being developed to combat the issues of decoherence and operational errors which are inherent in current quantum technologies.

> The development of scalable quantum technologies capable of accurate chemical simulations could drastically reduce the time and resources required for drug discovery, materials science, and the study of catalytic processes.

Furthermore, the use of entanglement and superposition in quantum systems offers a novel approach to exploring chemical dynamics and reaction mechanisms. By mapping the states of a molecule onto a quantum system, researchers can observe the evolution directly under different conditions and perturbations, leading to a more nuanced understanding of dynamic processes at the quantum level.

System	Classical Computation Time	Quantum Computation Time
Small molecule	Hours	Minutes
Medium molecule	Days	Hours
Large complex molecule	Infeasible	Days

Table 9.2: Comparison of computation times for classical and quantum systems based on molecule size.

In summary, quantum computing holds a substantial promise for the future of chemical simulations. By leveraging quantum mechanical properties, these advanced computational machines are poised to overcome the limitations of traditional methods, providing a new paradigm in the representation and understanding of chemical systems.

9.6 High-Performance Computing in Quantum Chemistry

High-Performance Computing (HPC) has transformed quantum chemistry by enabling the execution of complex simulations and calculations that were previously unfeasible due to computational limits. This section delves into the pivotal role of HPC in the field, elucidating its contributions to algorithm optimization, scalability of simulations, and the handling of larger, more complex molecular systems.

HPC systems are composed of interconnected compute nodes that often include thousands of processors and graphics processing units (GPUs) working in parallel. This architecture significantly reduces the time required for performing large-scale quantum mechanical calculations, which are inherently parallel in nature. The main components of HPC in quantum chemistry include parallelization of computational algorithms, optimization of electronic structure calcula-

9.6. HIGH-PERFORMANCE COMPUTING IN QUANTUM CHEMISTRY

tions, and effective data handling and storage solutions.

Parallelization of Computational Algorithms Quantum chemistry algorithms benefit markedly from parallelization, where the computational load is distributed across multiple processors. Schrödinger's equation, at the heart of quantum mechanics, requires solving multi-dimensional integral and differential equations. For instance, methods like Hartree-Fock and post-Hartree-Fock incorporate iterative solutions that lend themselves well to parallel execution.

In parallel algorithm design, tasks are divided into smaller, independent tasks that can be executed simultaneously. This not only speeds up computations but also allows for the treatment of larger molecular systems. For example, the parallel implementation of the Coupled Cluster (CC) singles and doubles method, which has high memory and computational demands, can be scaled efficiently using distributed memory parallel computing strategies.

Optimization of Electronic Structure Calculations The quantum chemical landscape frequently employs methodologies such as Density Functional Theory (DFT) and Wavefunction-based methods. HPC has catalyzed the development of optimized software packages that handle the complexities of electronic structure calculations with improved efficiency. Quantum chemistry software typically utilizes optimized mathematical libraries and hardware-specific tuning to enhance performance.

For instance, the utilization of advanced vectorization techniques and the exploitation of cache memory optimizations are common in HPC applications to minimize data movement and maximize computational speed. Furthermore, recent developments include hybrid approaches that use both CPUs and GPUs to synergistically speed calculations, combining the strong point of each technology.

Effective Data Handling and Storage Solutions Handling the vast amounts of data generated by quantum chemical simulations is a critical aspect of HPC. Efficient storage, retrieval, and processing of this data are paramount for productive analysis and interpretation. Techniques such as hierarchical data formats (e.g., HDF5) and parallel file systems like Lustre or GPFS enable efficient data management strategies.

Additionally, the integration of cloud computing with HPC environments is beginning to alter the landscape. Cloud-based HPC offers scalability and flexibility, allowing researchers to dynamically allocate resources based on their computational needs. This is particularly advantageous for ad-hoc projects or for small research groups without access to large in-house HPC facilities.

The potent combination of advanced algorithms, optimized computational methodologies, and robust data handling mechanisms not only makes HPC an indispensable tool in quantum chemistry but also continuously expands the horizon of what can be achieved in this vibrant field of study. Whether it is refining the accuracy of predictive models, exploring new materials, or understanding complex biological processes, HPC stands as a cornerstone in the advancement of quantum chemistry.

9.7 Multi-scale Modeling Techniques

Multi-scale modeling techniques are imperative in bridging the gap between quantum mechanical descriptions at the atomic scale and macroscopic phenomena observed in complex systems. These methods integrate different scales of modeling, from quantum mechanics at the smallest scale, through molecular mechanics, to continuum models. Each modeling level brings unique insights, and their integration allows for the accurate prediction of material behavior under various conditions.

At the quantum level, the principal method employed is the Density Functional Theory (DFT), which provides detailed information about the electronic structure of atoms and molecules. DFT calculations, however, are computationally intense and generally limited to systems containing a few hundred atoms. When considering larger systems, such as biomolecules or materials, a purely quantum mechanical treatment becomes impractical due to the exponential growth of computational requirements.

To address larger systems, multi-scale techniques employ hierarchical modeling which combines quantum mechanics (QM) with molecular mechanics (MM). Such QM/MM methods treat the region of interest, where electronic phenomena play a crucial role, with quantum mechanics, while the remainder of the system is modeled using

9.7. MULTI-SCALE MODELING TECHNIQUES

classical mechanics which requires significantly less computational resource. This hybrid approach makes it feasible to study large biological molecules like proteins, enzymes, and nucleic acids, where only specific parts of the molecule, such as active sites, require the precision of quantum mechanics.

The *partitioning* of the system in QM/MM methods can be based on different criteria, including geometric selection, where the atoms within a certain radius of a site of interest are treated quantum mechanically, or based on electronic criteria, where atoms are selected for their impact on the electronic density of the system. This partitioning is crucial for reducing the computational load while maintaining accuracy where it is most needed.

The partitioning leads to border issues between the QM and MM regions, often handled through various boundary schemes. One common approach is the use of link atoms, typically hydrogen atoms, that cap dangling bonds at the boundary, enabling smoother QM/MM interactions.

Moreover, the advent of high-performance computing has facilitated the deployment of more encompassant multi-scale models, incorporating not only QM and MM but also mesoscale and continuum modeling approaches. This is particularly relevant in materials science, where properties such as fracture toughness, thermal conductivity, and macroscopic deformations are crucial and require understanding that spans multiple length and time scales.

In computational practice, the handling of such multi-scale models involves careful coordination between different software packages that specialize in each scale, with data passing between them either sequentially or iteratively. This requires robust and well-defined interfaces as well as accurate translation of data. Considerations include the consistency of interatomic potentials used in quantum and classical regions, and the alignment of thermodynamic conditions across scales.

Recent advancements have seen the introduction of automated multi-scale modeling frameworks that aim to simplify the process of setting up and running these complex simulations. These frameworks often come with pre-built scenarios for common multi-scale problems, reducing the barrier for new researchers in the field.

The evolution of multi-scale modeling continues to be a vital aspect of quantum chemistry, particularly as new computational methods and hardware become available. These advances enable more realistic simulations of complex systems that are closer to experimental conditions, thereby providing deeper insights and more accurate predictions.

Overall, multi-scale modeling techniques serve as a cornerstone in the understanding and prediction of physical and chemical properties of complex systems, which is invaluable in drug design, material science, and beyond. These techniques, although computationally challenging and intricate in their setup, offer a comprehensive toolset for bridging the microscopic world of electrons and atoms with the macroscopic world we observe directly.

9.8 Time-Dependent Dynamics Simulations

Time-dependent dynamics simulations in quantum chemistry are critical for understanding the evolution of molecular systems over time, which is essential for fields such as photodynamics, catalysis, and materials science. The fundamental theory underlying these simulations originates from the time-dependent Schrödinger equation (TDSE), which describes how the quantum state of a physical system changes over time.

The general form of the TDSE is:

$$i\hbar \frac{\partial}{\partial t} \Psi(\mathbf{r}, t) = \hat{H} \Psi(\mathbf{r}, t)$$

where i is the imaginary unit, $\hbar$ is the reduced Planck's constant, $\Psi(\mathbf{r}, t)$ is the wave function of the system, dependent on position $\mathbf{r}$ and time t, and $\hat{H}$ represents the Hamiltonian operator.

One major challenge in solving the TDSE for complex molecular systems is its high computational demand, as it involves the evolution of a multi-dimensional wave function over time. To address these challenges, various approximate and numerical methods have been developed.

Time-dependent Density Functional Theory (TDDFT) is one such approach, extending the principles of ground-state Density Functional Theory to dynamic properties and excitations. TDDFT simplifies the problem by using time-dependent density $\rho(\mathbf{r}, t)$ instead

9.8. TIME-DEPENDENT DYNAMICS SIMULATIONS

of the wave function. The key equation in TDDFT is:

$$i\frac{\partial}{\partial t}\rho(\mathbf{r},t) = [\hat{H}, \rho(\mathbf{r},t)]$$

Another widely used method is the **Time-dependent Hartree-Fock (TDHF)** theory, where the Fock matrix is time-dependent, and its solution provides information about the system's dynamical properties.

Quantum dynamics simulations often employ numerical techniques such as:

- *Split-operator techniques*, which separate the kinetic and potential energy terms in the Hamiltonian for independent treatment during small time steps.

- *The Crank-Nicolson method*, a finite difference method used in numerical solutions of the TDSE by discretizing time and space coordinates, given by:

$$(\mathbf{I} + \frac{i\Delta t}{2\hbar}\hat{H})\Psi(\mathbf{r}, t + \Delta t) = (\mathbf{I} - \frac{i\Delta t}{2\hbar}\hat{H})\Psi(\mathbf{r}, t)$$

where $\mathbf{I}$ is the identity matrix, and Δt is a small time increment.

- *Chebyshev polynomial methods*, which efficiently handle the propagation of wave functions over longer times by approximating the exponential operator in the time evolution operator $\exp(-i\hat{H}t/\hbar)$.

Furthermore, recent developments include machine learning algorithms to predict the evolution of molecular systems, significantly reducing the computational cost by learning from precomputed databases of dynamic simulations.

By integrating these various methods into simulations, significant advancements have been made in predicting reaction dynamics, excited state dynamics, electron transfer processes, and more in a time-dependent framework. These tools allow chemists to explore potential energy surfaces and transition states with greater accuracy than was previously possible.

While the computational complexity and cost associated with time-dependent dynamics simulations remain high, ongoing developments in numerical methods, algorithms, and computational resources continue to advance our ability to predict and understand the dynamic behavior of molecular systems with high precision and speed.

9.9 New Functional and Basis Sets

The recent development of new functional and basis sets in quantum chemistry marks significant advancements that enhance the predictive power and efficiency of calculations. Functional in the context of Density Functional Theory (DFT) represents the mathematical expressions that model the electron density of a molecular system. Basis sets, on the other hand, are sets of functions used to describe the electronic wave functions of atoms and molecules within molecular orbital theory.

Advancements in Functional Development

In DFT, the accuracy of the results heavily depends on the choice of functional, which includes the exchange-correlation functional. Recent years have seen the development of more accurate functionals that attempt to overcome the limitations of popular generalized gradient approximation (GGA) and hybrid functionals. One notable development is the increased use of meta-GGA functionals, which incorporate additional variables like the kinetic energy density of electrons to improve interaction modeling.

A specific example of an advanced functional is the SCAN (Strongly Constrained and Appropriately Normed) functional, which has demonstrated superior performance in computing properties of molecules and solids that challenge traditional functionals. The mathematical representation of the SCAN functional is given by:

$$E_{xc}[n] = \int \epsilon_{xc}(\rho(\mathbf{r}), \nabla\rho(\mathbf{r}), \tau(\mathbf{r})) \, d^3r,$$

where $\rho(\mathbf{r})$ is the electron density, $\nabla\rho(\mathbf{r})$ is its gradient, and $\tau(\mathbf{r})$ is the kinetic energy density.

Development in Basis Sets

Parallel to the improvements in functionals, there has been substantial progress in the development of more comprehensive and efficient basis sets. Basis sets aim to provide a balanced description of electron interactions within a molecule. Recent trends include the creation of basis sets that reduce computational cost without sacrificing accuracy. An example of such a basis set is the def2 family of basis sets, which includes polarization functions and is optimized for use with DFT calculations.

These basis sets often provide a better compromise between computational speed and accuracy, and their use is illustrated in the following example. Consider a DFT calculation on a simple molecule, methane (CH_4), using the def2-SVP basis set. The basis set notation implies:

- "def2": Indicates the family of continuously improved basis sets by the developers.
- "SVP": Stands for Split Valence with Polarization, where the valence electrons are described by different numbers of basis functions, and polarization functions are added to improve accuracy.

The inclusion of polarization functions is crucial as these functions allow for anisotropic electron density distribution, enhancing the modeling of molecular distortion and interaction. The computational representation uses basis functions like Gaussian orbitals, represented as:
$$\chi_i(\mathbf{r}) = (x - X_A)^l (y - Y_A)^m (z - Z_A)^n e^{-\alpha_i (\mathbf{r} - \mathbf{R}_A)^2},$$
where (X_A, Y_A, Z_A) are the coordinates of atom A, and α_i are the Gaussian parameters that define the width of the orbital.

The development of new functionals and basis sets in recent years significantly enhances the scope and accuracy of quantum chemical calculations. These advancements allow for a more refined exploration of molecular systems, from simple molecules to complex molecular assemblies, paving the way for deeper insights into chemical reactions and properties.

9.10 Green Chemistry and Sustainable Approaches

In the context of quantum chemistry, the integration of green chemistry principles signifies a critical shift toward sustainable scientific practices that both prevent waste and reduce the environmental footprint of chemical processes. Quantum chemical methodologies can be pivotal in predicting and optimizing reactions that align with these principles by focusing on atom economy, less hazardous chemical synthesis, and energy efficiency.

The fundamental tenet of green chemistry is to design chemical products and processes that reduce or eliminate the use and generation of hazardous substances. Quantum chemistry contributes to this goal by enabling the simulation of chemical reactions at the molecular level, which allows chemists to predict outcomes of reactions and the environmental impact of various substances without the need for extensive physical experiments.

An increasingly significant role is played by Density Functional Theory (DFT) due to its ability to provide insights into the electronic structure of molecules, which is crucial for understanding and predicting chemical reactivity and properties. DFT calculations, for example, can be used to predict catalysts that are more efficient, selective, and capable of operating under milder conditions, thus reducing energy consumption and by-product formation.

To illustrate, consider the catalytic reduction of carbon dioxide, a vital reaction for addressing the greenhouse gas accumulations in the atmosphere. Advanced DFT methods can be used to model and predict catalysts that can selectively and efficiently convert CO_2 into useful hydrocarbons. This capability not only provides ways to mitigate environmental impact but also enhances the feasibility of recycling CO_2 into valuable products, a key principle in sustainable and green chemistry.

Machine learning techniques also intersect significantly with quantum chemistry for sustainable approaches. By employing machine learning algorithms to quantum chemical data, one can identify patterns and predict outcomes much faster than traditional computational chemistry approaches. For instance, machine learning models trained on large datasets of molecular structures and their properties can predict the most effective catalysts and optimal conditions for a reaction, substantially reducing the materials and energy required for experimental investigations.

Quantum computing holds further promise by potentially solving complex molecular systems that are currently infeasible for classical computers. This capability could revolutionize how we design drugs, materials, and chemicals, offering pathways to unprecedented accuracy in simulations and drastic reductions in trial and error experimental methods.

Moreover, multiscale modeling techniques bridge the gap between quantum mechanical calculations at the atomic or molecular scale and larger scale phenomena. These techniques, by encompassing different levels of detail, provide a comprehensive understanding that can lead to the design of more effective and less polluting systems. For example, coupling quantum mechanical methods with classical statistical mechanics can optimize the efficiency of photovoltaic cells, contributing to more sustainable energy sources.

In the practical realm, application of such advanced quantum chemical methods aligns closely with the 12 principles of green chemistry. These include reducing energy demand by identifying the least energy-intensive pathways, developing renewable materials, designing safer molecules, and improving the selectivity of reactions.

To encapsulate, the intersection of quantum chemistry with green chemistry principles provides a powerful toolkit for advancing chemical research towards more sustainable and environmentally friendly outcomes. Through advanced simulations and computational chemistry, researchers are not only able to enhance the understanding of complex chemical systems but also drive significant shifts in traditional chemical synthesis towards greener alternatives that are less resource-intensive and more ecologically benign.

9.11 Quantum Chemistry in Nanotechnology and Nanomedicine

The integration of quantum chemistry into nanotechnology and nanomedicine has been pivotal in driving innovations that enhance molecular functionality, targeted drug delivery, and diagnostic methods at the nanoscale. This section explores the roles quantum chemistry plays within these interdisciplinary fields, emphasizing how quantification of electronic, atomic, and molecular interactions can lead to a deeper understanding and improved applications.

Quantum chemistry methodologies facilitate the precise simulation of nanomaterials, which are critical for designing drug delivery systems and medical diagnostic tools. Concepts such as molecular orbital theory, electron density properties, and potential energy surfaces become essential tools for predicting the behavior and interaction of molecules at the nanoscale. The Schrödinger equation serves as the cornerstone in describing the electronic structure of both single molecules and molecular assemblies in nanotechnology.

Molecular Design and Drug Delivery: In nanomedicine, drug delivery systems leveraging nanoparticles as carriers for therapeutic agents have shown benefits such as enhanced solubility, targeted delivery, and controlled release. Using quantum chemistry, researchers calculate the most stable molecular structures and potential reaction pathways to optimize the therapeutic efficacy and minimize unintended interactions of nanoparticles used in these systems.

For example, let's consider the design of a nanoparticle-based system for delivering cancer drugs. Quantum chemical simulations can identify the optimal size and functionalization of nanoparticles to maximize penetration through cellular membranes while avoiding detection by the immune system. These computational predictions rely on molecular dynamics simulations and density functional theory (DFT) to offer insight into the interactions between drug molecule and nanoparticle surface, which is often coated with targeting ligands that direct the drug to specific cancer cells.

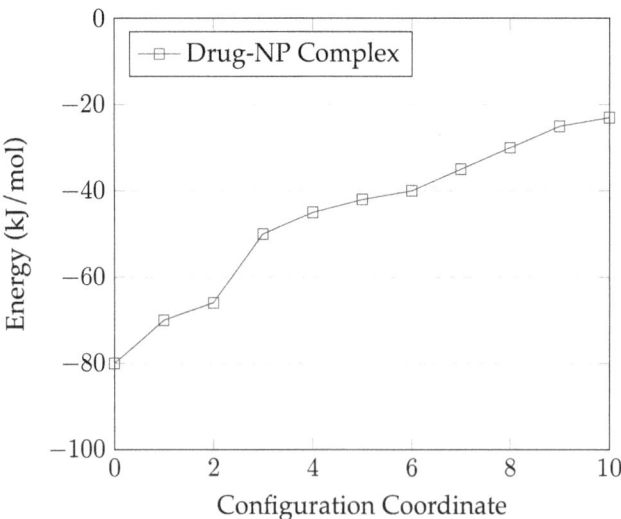

Potential Energy Surface for a Drug-Nanoparticle Complex

Diagnostic Techniques: Quantum chemical calculations are indispensable for designing fluorescent molecules used as probes in biological imaging. The electronic properties of these molecules, like excitation energies, can be tuned precisely to improve their fluorescent characteristics. By considering the interaction of light with matter fundamentally through quantum mechanics, researchers can predict and enhance the fluorescence quantum yield and stability of these probes under physiological conditions.

Additionally, the development of quantum dots for use in imaging and as fluorescent markers in cell tracking benefits significantly from theoretical predictions of electronic structure and band gap engineering. Quantum dots' customizable optical and electronic properties allow for a high degree of control over their spectral characteristics, making them highly effective for multiplexed imaging applications in biological systems.

Table 1: Comparison of Conventional and Quantum Dot Fluorescence Characteristics

Characteristic	Conventional Dye	Quantum Dot
Fluorescence Lifetime	Short	Extended
Photostability	Poor	Excellent
Emission Tunability	Limited	Wide Range

Quantum chemistry, by predicting and demonstrating how atoms and molecules assemble and interact at the nano-bio interface, fundamentally supports the design and optimization of nanosystems for therapeutic and diagnostic purposes in medicine. It facilitates a multi-scale understanding that transcends classical intuit and unveils the nuanced quantum mechanical effects driving the behavior of complex molecular systems at nanoscale dimensions. The increase in computational horsepower and the refinement of simulation techniques will continue to advance the capabilities and applications of quantum chemistry in nanotechnology and nanomedicine, promising continued developments in healthcare technologies.

www.ingramcontent.com/pod-product-compliance
Lightning Source LLC
Chambersburg PA
CBHW052149220526
45471CB00004B/1598